Solution of
Partial Differential Equations
on Vector and Parallel
Computers

James M. Ortega

Department of Applied Mathematics and Computer Science
University of Virginia
Charlottesville, Virginia

Robert G. Voigt

Institute for Computer Applications in Science and
Engineering
NASA Langley Research Center
Hampton, Virginia

siam

Philadelphia
1985

Library of Congress Catalog Card Number 85-61387
ISBN 0-89871-055-3

The preparation of this manuscript was supported by the U.S. Army Research Office under contract DAAG29-80-C-0091. James M. Ortega's research was supported in part by the National Aeronautics and Space Administration under NASA grant NAG-1-46. Robert G. Voigt's research was supported by the National Aeronautics and Space Administration under NASA contracts NASI-17070 and NASI-17130 while he was in residence at ICASE.

PREFACE

Numerical simulation of complex physical phenomena on digital computers has been the impetus behind the remarkable improvements in performance of such devices since their inception. Most of this improvement has come from faster and faster components, but in the last few years it has become increasingly difficult to achieve performance gains in this manner because of the barrier imposed by the speed of light. Consequently, computer designers have sought architectural techniques to help produce performance gains. One of the simplest of these conceptually, yet hardest to use effectively, is parallelism, the ability to compute on more than one part of the same problem at the same time.

In theory, parallelism solves all performance problems: if a faster computer is required for a given problem, one need only add a sufficient number of processors until the power of the aggregate matches the need. Of course, this is an extremely simplistic view, for as in any computing system, algorithms and software must be developed to take advantage of the power of the hardware.

In the 1960's and early 1970's, there was considerable research aimed at developing algorithms for systems that had such ideal characteristics as an arbitrarily large number of processors that could cooperate with zero overhead. By the mid-1970's, with the advent of vector computers that approximated parallelism by using a computational pipeline, interest shifted away from theoretical parallel computers and concentrated on developing methods to utilize machines that were actually available. Now in the 1980's, we are beginning to see computers with parallelism achieved by using a number of independent processors. Such systems should usher in an exciting period of algorithm research, since they have a variety of characteristics that are different from previous computers and that have a profound effect on algorithm performance.

The purpose of this volume is to review, in the context of partial differential equations, algorithm development that has been specifically aimed at computers that exhibited some form of parallelism. We will discuss architectural features of these computers insomuch as they influence algorithm performance, and hopefully, provide some insight into algorithm characteristics that allow effective utilization of the hardware.

We have attempted to compile a complete bibliography on the subject, up to the time of publication. But the literature in this field is expanding at a rapid rate. It is interesting to note that it was not until 1984 that the first journals devoted to parallel computing appeared, and perhaps because of this, much of the literature has appeared in the form of technical reports from government laboratories and universities.

Colleagues too numerous to mention have contributed to this review by providing references or commenting on the manuscript. Finally, we would like to acknowledge the efforts of Carolyn Duprey and Ann Turley, who prepared the text from our illegible scribbles, and Emily Todd, who prepared the bibliography knowing full well that "this is the last one" was a lie.

JAMES M. ORTEGA
ROBERT G. VOIGT
Charlottesville, Virginia
Hampton, Virginia
May, 1985

CONTENTS

Abstract. In this work we review the present status of numerical methods for partial differential equations on vector and parallel computers. A discussion of the relevant aspects of these computers and a brief review of their development is included, with particular attention paid to those characteristics that influence algorithm selection. Both direct and iterative methods are given for elliptic equations as well as explicit and implicit methods for initial-boundary value problems. The intent is to point out attractive methods as well as areas where this class of computer architecture cannot be fully utilized because of either hardware restrictions or the lack of adequate algorithms. A brief discussion of application areas utilizing these computers is included.

1. Introduction. For the past 20 years, there has been increasing interest in the use of computers with a parallel or pipeline architecture for the solution of very large scientific computing problems. As a result of the impending implementation of such computers, there was considerable activity in the mid and late 1960's in the development of parallel numerical methods. Some of this work is summarized in the classical review article of Miranker [1971]. It has only been in the period since then, however, that such machines have become available. The Illiac IV was put into operation at NASA's Ames Research Center in 1972; the first Texas Instruments Inc. Advanced Scientific Computer (TI-ASC) became operational in Europe in 1972; the first Control Data Corp. STAR-100 was delivered to Lawrence Livermore National Laboratory in 1974; and the first Cray Research Inc. Cray-1 was put into service at Los Alamos National Laboratory in 1976.

Since 1976, the STAR-100 has evolved into the CDC Cyber 203, which is no longer in production, and the Cyber 205, which is now CDC's entry in the supercomputer field. The Cray-1 has evolved into the Cray-1S, which has considerably more memory capability than the original Cray-1, and the Cray X-MP, a faster multiprocessor version. On the other hand, the TI-ASC is no longer in production, and the

Illiac IV ceased operation in 1981. For the last twenty years, the most expensive commercial computer at any given time has cost in the $10–$20 million dollar range; this is still the correct interval for today's supercomputers.

The Illiac IV consisted of 64 processors. Other computers consisting of a (potentially large) number of processors include the Denelcor HEP and the International Computers Ltd. DAP, both of which are offered commercially, and a number of one of a kind systems in various stages of completion or development: the Finite Element Machine at NASA's Langley Research Center; MIDAS at the Lawrence Berkeley Laboratory; Cosmic Cube at the California Institute of Technology; TRAC at the University of Texas; Cm* at Carnegie–Mellon University; ZMOB at the University of Maryland; Pringle at the University of Washington and Purdue University; and the MPP at NASA's Goddard Space Flight Center. The first two of the latter group of machines are designed primarily for numerical computation while the others are for research in computer science, for image processing, etc. A recent development made possible by the increasing power and flexibility of microprocessors and the dropping cost of fabrication is the emergence of several small entrepreneurial companies offering commercial parallel and vector systems at modest prices. Examples include Elxsi, Flexible Computer, Inc. and Convex, Inc.

Other computers of some historical interest, although their primary purpose was not for numerical computation, include Goodyear Corporation's STARAN (Goodyear [1974], Gilmore [1971], Rudolph [1972], and Batcher [1974]), and the C.mmp system at Carnegie–Mellon University (Wulf and Bell [1972]). Also of some historical interest, although it was not brought to the market, is Burroughs Corporation's Burroughs Scientific Processor (Kuck and Stokes [1982]).

During the last 15 years, the literature on parallel computing has been increasing at a rapid rate and a number of books and survey papers have been written which complement the present work. The book by Hockney and Jesshope [1981] contains much information on architectures as well as languages and numerical methods. Other books or surveys dealing with architecture or other computer science issues or applications include Worlton [1981] and Zakharov [1984] on the history of (and future for) parallel computing, Hord [1982] on the Illiac IV, Kogge [1981] on pipelining, Avizienis. et al. [1977] on fault-tolerant architectures for numerical computing, Hockney [1977], Kuck [1977], [1978], Kung [1980], Stone [1980] and Uhr [1984]. Surveys on numerical methods include, in addition to Miranker [1971] already mentioned, Traub [1974a], Poole and Voigt [1974], which was an essentially complete annotated bibliography up to the time of its publication, Heller [1978], which concentrates on linear algebra problems and gives considerable attention to theoretical questions, T. Jordan [1979], which summarizes performance data for linear algebra software for several vector computers of the late 1970's, Book [1981], Buzbee [1981], Evans [1982a], which also contains a number of nonnumerical articles, Sameh [1977], [1981], [1983], Voigt [1977], Ortega and Voigt [1977] which the present work updates, Rodrigue [1982], a collection of review papers on various numerical methods and applications, Gentzsch [1984b], which concentrates on vectorization of algorithms for fluid mechanics, and Schnendel [1984], an introductory textbook.

There are also several interesting papers which review the need and uses for supercomputers. These include Ballhaus [1984], Buzbee [1984a], Buzbee, et al. [1980], Chapman [1979], Fichtner, et al. [1984], Gautzsch, et al. [1980], Gloudeman [1984], Hockney [1979], Inouye [1977], Kendall, et al. [1984], Lomax [1981], Peterson [1984a, b], Rodrigue, et al. [1980], and Williamson and Swarztrauber [1984]. Finally, we mention that there has been increasing interest in the use of add-on array processors such as those made by Floating Point Systems, Inc. (Floating Point Systems [1976]), but this topic is beyond the scope of this paper; see, for example, Eisenstat and Schultz [1981] and Wilson [1982].

The challenge for the numerical analyst using vector or parallel machines is to devise algorithms and arrange the computations so that the architectural features of a particular machine are fully utilized. As we will see, some of the best sequential algorithms turn out to be unsatisfactory and need to be modified or even discarded. On the other hand, many older algorithms which had been found to be less than optimal on sequential machines have had a rejuvenation because of their parallel properties. In §§3 and 4 we review the current state of parallel algorithms for partial differential equations, especially elliptic boundary value problems. In §3 we discuss direct methods for the solution of linear algebraic systems of equations while in §4 we consider iterative methods for linear systems as well as time-marching methods for initial and initial-boundary value problems. Finally, in §5, we briefly review selected applications which have been reported in the literature.

In order to have a framework in which to study and evaluate algorithms, a variety of concepts have been introduced which we will use in the algorithm discussions that follow. Many of these ideas are becoming widely accepted as a basis for study and we introduce them in general terms now.

Traditionally, one of the most important tools of the numerical analyst for evaluating algorithms has been computational complexity analysis, i.e., operation counts. The fact that the fast Fourier transform of n samples requires $O(n \log n)$ arithmetic operations (here and throughout, log denotes $\log_2$) while the straightforward approach requires $O(n^2)$ provides a clear choice of algorithms for serial computers. This arithmetic complexity remains important for vector and parallel computers, but several other factors become equally significant. As we will see in the next section, vector computers achieve their speed by using an arithmetic unit that breaks a simple operation, such as a multiply, into several subtasks, which are executed in an assembly line fashion on different operands. Such so-called vector operations have an overhead associated with them that is called the start-up time, and vector operations are faster than scalar operations only when the length of the vector is sufficient to offset the cost of the start-up time. In §3, we show that this start-up time typically enters the complexity formula as a coefficient of the next to the highest order term. Thus, terms that are neglected in the usual complexity analysis may play a prominent role in choosing algorithms for vector computers.

Nor is it sufficient just to minimize the number of vector operations. Every arithmetic operation costs some unit of time on a vector computer even if it is part of a vector operation. Thus, for vectors of length n, an algorithm that requires $\log n$ vector operations will not be faster for sufficiently large n than an algorithm that requires n scalar operations since $n \log n$ operations will be performed. This preservation of arithmetic complexity is made more precise by the introduction of the concept of consistency in §3, and we will show that in general for vector computers one should choose algorithms whose arithmetic complexity is "consistent" with the best scalar algorithm.

Two techniques for improving the performance of vector computers involve the restructuring of DO loops in Fortran in order to force a compiler to generate an instruction sequence that will improve performance. It is important to note that the underlying numerical algorithm remains the same. The technique of rearranging nested

DO loops is done to help the compiler generate vector instructions. For example,

$$DO\ 100\ I = 1, N$$
$$DO\ 100\ J = 1, N$$
$$100\qquad B(I) = B(I) + A(I, J)$$

would yield scalar add instructions and would be changed to

$$DO\ 100\ J = 1, N$$
$$DO\ 100\ I = 1, N$$
$$100\qquad B(I) = B(I) + A(I, J)$$

resulting in a vector add instruction for each value of J. The other technique, characterized as unrolling DO loops in Dongarra and Hinds [1979], is used as a way to force the compiler to make optimal use of the vector registers on the Cray computers. (The role of these registers will be discussed in the next section.) In its simplest form, loop unrolling involves writing consecutive instances of a DO loop explicitly with appropriate changes in the loop counter to avoid duplicate computation. Several examples are given by Dongarra [1983] and Dongarra and Eisenstat [1984] for basic linear algebra algorithms. Although of little value in helping to evaluate different numerical algorithms, these techniques do provide insight into how to obtain maximum performance on vector computers.

The previous two examples indicate some of the limitations with present Fortran compilers, but a general discussion of compilers for vector and parallel computers, though crucial to performance, is beyond the scope of this review. For discussions of the present state of the art see, for example, Arnold [1982], [1983], Kuck, McGraw and Wolfe [1984], and Kuck, et al. [1984].

The above discussion has focused on vector computers, and although some of the issues are relevant to computers consisting of parallel processors, there are other important considerations as well. Arithmetic complexity remains fundamental but extra computations may not involve the penalty that they would on vector computers (if, for example, there are processors that would otherwise be idle). Equally important will be the degree of parallelism, the amount of the computation that can be done in parallel, which will be defined in §3 and used extensively in the discussions on algorithms. We will see that there are algorithms with relatively high operation counts that are attractive on parallel computers because a high percentage of those operations can be done in parallel.

As emphasized by Gentleman [1978], a nonnumerical issue that is crucial to the performance of algorithms on parallel computers is the frequency and cost both of communication among processors and of synchronization of those processors. A simple iterative method provides an example. If unknowns are distributed among processors and if the new approximate solution has been computed in these processors, then parts of this solution must be communicated to other processors in order to compute the next iterate. The amount and destination of this information depends on the underlying problem, on how it is mapped onto the processors, and on the numerical algorithm. Once the communication takes place there must be synchronization if the processors are to stay on the same iteration step. There are a number of ways to do this, with varying costs depending on the architecture. Many examples of communication and synchronization costs will be brought out in §§3 and 4 but they will not be incorporated into a formal complexity analysis. Such analyses are only beginning to appear and a more complete discussion of the costs and how to analyze them may be found in Adams and Crockett [1984], Reed and Patrick [1984a, b] and Gannon and Van Rosendale [1984b].

Less formal consideration of communication and synchronization involves assumptions such as an equal cost to communicate one floating point number and to perform one floating point operation. As an extreme case, one can assume zero cost to communicate. This zero-cost model, although unrealistic, can provide useful bounds on the performance of an algorithm, and it was this motivation that led to the proposal of the Paracomputer by Schwartz [1980]. In this model the parallel array contains an unbounded number of processors all of which may access a common memory with no conflict and at no cost. Such unrestrained resources make it possible to study the inherent, total parallelism in an algorithm and to obtain an indication of its optimal performance. It also provides a standard by which to measure the effectiveness of other architectures. Some of the algorithm development discussed in this review fits the paracomputer model. The paracomputer assumption of an unbounded number of processors has historically been a popular assumption and Heller [1978] reviews research of this kind, particularly for linear algebra algorithms.

At the opposite end of the spectrum from the paracomputer are actual running arrays where the number of processors is obviously fixed, and (for the immediate future) usually small relative to the size of the problem. These systems motivate research on models involving p processors where p is fixed and is much less than n, a parameter measuring the size of the problem. In between, one finds the model of a system with the number of processors given as some simple function of n. We will see that these different models can lead to different algorithms for the same problem.

Most parallel numerical algorithms follow one or both of two related principles which we refer to as *divide and conquer* and *reordering*. The divide and conquer approach involves breaking a problem up into smaller subproblems which may be treated independently. Frequently, the degree of independence is a measure of the effectiveness of the algorithm for it determines the amount and frequency of communication and synchronization. Applying the divide and conquer concept to the inner product computation $\Sigma a_i b_i$, where the product $a_i b_i$ has been computed in processor p_i, might involve sending $a_{i+1} b_{i+1}$ to processor p_i, for i odd. The sum operation is now "divided" among $p/2$ processors with p_i doing the addition $a_i b_i + a_{i+1} b_{i+1}$ for i odd. The idea is repeated $\log n$ times until the sum is "conquered" in processor p_1. There are several other ways to organize the computation, all of which will be superior (on reasonable architectures) to simply sending all the products $a_i b_i$ to a single processor for summation. We will see that this simple idea pervades many parallel algorithms.

The concept of reordering may be viewed as restructuring the computational domain and/or the sequence of operations in order to increase the percentage of the computation that can be done in parallel. For example, the order in which the nodes of a grid are numbered may increase or decrease the parallelism of the algorithm to be used. An analogous example is the reordering of the rows and columns of a matrix to create independent submatrices that may be processed in parallel. Specific algorithms based on this concept will be discussed in §§3 and 4.

After one has obtained a parallel algorithm it is natural to try to measure its performance in some way. The most commonly accepted measure is speedup, which is frequently defined as

$$S_p = \frac{\text{execution time using one processor}}{\text{execution time using } p \text{ processors}}.$$

The strength of this definition is that it uses execution time and thus incorporates any communication or synchronization overhead. A weakness is that it can be misleading to focus on *algorithm* speedup when in fact one is usually more interested in how much faster a problem can be solved with p processors. Thus, we wish to compare the best serial algorithm with the parallel algorithm under consideration, and we define

$$S'_p = \frac{\text{execution time using the fastest sequential algorithm on one processor}}{\text{execution time using the parallel algorithm on } p \text{ processors}}.$$

This second definition makes clear that an algorithm with excellent parallel characteristics, that is, a high speedup factor S_p, still might not yield as much actual improvement on p processors as S_p would indicate.

Ware [1973] suggested another definition of speedup in order to reflect more clearly the role of scalar computation in a parallel algorithm:

$$S_p = \left[(1-\alpha) + \frac{\alpha}{p}\right]^{-1}.$$

Here α is the fraction of work in the algorithm that can be processed in parallel, and the execution time using a single processor has been normalized to unity. Buzbee [1983c] points out that

$$\left.\frac{dS_p}{d\alpha}\right|_{\alpha=1} = p^2 - p$$

and this quadratic behavior is shown in Fig. 1.1 where it is clear that the fraction of work that can be done in parallel must be high to achieve reasonable speedups. Buzbee also points out the similarity between Fig. 1.1 and the behavior of vector performance if the abscissa is interpreted as the fraction of vectorizable work. Buzbee [1983b] uses the Ware model to discuss the parallel properties of particle-in-cell codes for fusion studies, concluding that a large percentage of the work can be processed in parallel.

Buzbee [1983c] also notes that a weakness of the Ware model of speedup is that $S_p = p$ for an algorithm that is completely parallel ($\alpha = 1$), which is unlikely because of various overheads associated with parallel computation. In fact, Minsky [1970] conjectured that speedup for p processors would be proportional to $\log p$. Buzbee suggests the

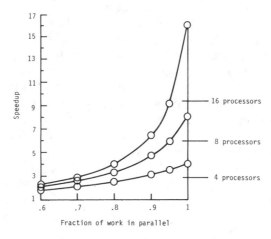

FIG. 1.1. *Speedup as a function of parallelism and number of processors.*

following change to the Ware model:

$$S_p = \left[(1-\alpha) + \frac{\alpha}{p} + \sigma(p) \right]^{-1}$$

where $\sigma(p)$ reflects the overhead of using p processors. Thus it is clear that if we are to improve on the Minsky conjecture, algorithms and implementations must be found for which α is near unity and $\sigma(p)$ is near zero. Other studies by Kuck, et al. [1973] and Lee [1977] suggest that over a broad range of problems it is reasonable to expect an average speedup proportional to $p/\log p$.

Knowing the speedup, it is reasonable to ask how efficiently the parallel system is being utilized by the algorithm. One way to accomplish this is to use the efficiency measure defined by

$$E_p \equiv \frac{S_p}{p}.$$

Thus in the ideal situation of a speedup of p for p processors, the efficiency measure is unity. For some other speedup factors, such as the conjectured $p/\log p$ discussed above, E_p tends to zero as p is increased, giving a clear indication that certain algorithms may not yield good processor efficiency for systems with a large number of processors.

In §2, we will review in more detail various architectural features of both pipelined computers and arrays of processors, and give further details on some of the machines mentioned in this section, as well as others. Among the topics that will not be discussed in §2 are digital optical computing and special devices designed to execute a specific algorithm or to solve a specific problem. Digital optical computing utilizes photons as the information carrying media, but, generally, issues involving algorithms are the same as for conventional digital computing. For a review see Sawchuk and Strand [1984] and for a discussion of some algorithmic considerations see Casasent [1984]. Computers designed specifically for an algorithm or problem are receiving increased attention because of the dropping cost of components. One such system, described by Christ and Terrano [1984], would deliver several billion floating point operations per second for elementary particle physics calculations.

2. Review of the hardware. In this section we shall review some of the basic features of vector and parallel computers. However, because of the plethora of such systems, each differing in detail from the others, we shall attempt to stress some of the basic underlying architectural features, especially as they affect numerical algorithms, and refer the reader to the literature for more details on the individual computers. Another reason that we shall not attempt to give a detailed treatment of any particular system is that the field is changing so rapidly. For example, as of this writing, Cray Research Inc. has announced the Cray-2, Control Data Corp. the Cyberplus, Denelcor the HEP-2, ETA the GF-10, and there are a number of university development projects. Moreover, there is an expected impact from VLSI technology, although the precise form this will take is not yet clear. Finally, the Japanese are developing several systems (Buzbee, et al. [1982], Kashiwagi [1984] and Riganati and Schneck [1984]).

One obvious way to achieve greater computational power is to use faster and faster circuits, and improvements in this area have been immense. However, the limits on transmission speeds imposed by the speed of light and fabrication limitations (see, for example, Seitz and Matisoo [1984]) have led to attempts to improve performance by parallelism, which, in its most general form, occurs whenever more than one function is being performed at the same time. This idea actually dates back to the ENIAC, the first electronic computer (Eckert, et al. [1945]), which was capable of executing many arithmetic operations simultaneously. However, the authors discussed several levels of parallelism and concluded that serial operation was to be preferred. One of their reasons—the difficulty of programming for parallel operations—was certainly prophetic. They also observed that improving component speeds made parallelism unnecessary!

Parallelism did reappear occasionally in various forms beginning in the early 1950's. For example, there was parallelism in the arithmetic units of the Institute for Advanced Study computer at Princeton University and The Whirlwind I at the Massachusetts Institute of Technology (Kuck [1978]), and parallelism between program execution and I/O on the UNIVAC I (Kogge [1981]). For a brief history of computers and an excellent guide to the literature the reader is referred to Kuck [1978].

The general notion of parallelism discussed above is basically the same as that set forth by Hobbs and Theis (Hobbs, et al. [1970]) but is too broad for our interests here since we are focused primarily on numerical algorithms. For our purposes, parallelism in computing will be exemplified by those computers which contain instructions for performing arithmetic operations on a collection of operands as an entity, such as vectors, or which contain independent processing elements that may be used on the arithmetic aspects of the same problem simultaneously.

Pipelining. One way of obtaining significant speedups is by the technique known as pipelining. We will use the term pipelining (as given in Kogge [1981]) to refer to design techniques that take some basic function to be invoked repeatedly and partition it into several subfunctions which can be done in an assembly line fashion. This is illustrated in Fig. 2.1, which shows how a floating point instruction is broken down into more elementary parts.

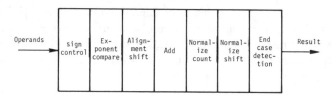

FIG. 2.1. *Floating point pipeline.*

By the early 1960's pipelining was being used in a variety of computers to speed up functions like memory access and instruction execution (Kogge [1981]). Eventually the technique was used in arithmetic units on the CDC 7600 and the IBM System 360 Model 91. However these computers do not fit our view of parallelism because the arithmetic instructions are executed with only one set of operands. The last necessary step was taken with computers such as the CDC Cyber 200 series (formerly the STAR-100), the Cray series and the TI-ASC, which have hardware instructions which accept vectors as operands. Since the ASC is no longer available, we will focus on the former two computers (see Watson [1972] for a description of the ASC). For simplicity, thoughout the remainder of this paper we will refer to the Cyber 200 series including the 203 and 205 as the Cyber 200 and to the Cray family as the Cray unless there is some reason to make a further distinction.

Cyber 200. We next give a very brief functional description of the Cyber 200 and Cray families. A thorough review of the Cray-1, the Cyber 205 and the Cray X-MP may be found in Russell [1978], Lincoln [1982] and Chen [1984], respectively; see also Larson [1984] for the X-MP. The Cyber 200 has separate arithmetic units for scalar and vector floating point arithmetic. The latter units, which we shall refer to as pipelines, are accessed by hardware vector instructions which obtain their operands directly from main memory. Main memory size ranges from 0.5 to 2 million 64-bit words on the 203 and 1 to 16 million on the 205 with further increases in memory size already announced. The 203 had two separate pipelines while the 205 may have 1,2 or 4. The pipelines are reconfigurable via microcode in order to execute a variety of arithmetic operations. A schematic of the Cyber 200 is given in Fig. 2.2.

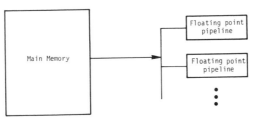

FIG. 2.2. *Cyber 200 schematic.*

A vector hardware instruction initiates the flow of operands to the pipeline, and assuming that the instruction involves two source vectors, each segment of the pipeline accepts two elements, performs its particular function (e.g., exponent adjustment), passes the result to the next segment, and receives the next two elements from the stream of operands. Thus, several pairs of operands are being processed concurrently in the pipeline, each pair in a different stage of the computation. The number of results emerging from the pipeline each clock period (cycle time) depends upon the arithmetic operation and the word length (64 bits or 32 bits). The maximum result speeds are given in Table 1 for various cases. For a 4-pipeline 205, the computation rates shown in Table 1 are doubled. Moreover, the 205 has the capability of handling "linked triads" of the form vector + constant × vector at the same rate as addition; hence, this operation achieves 200 million floating point operations per second (MFLOPS) for 64-bit arithmetic on a 2 pipeline machine, and 800 MFLOPS for 32-bit arithmetic on a 4 pipeline machine. This is the fastest possible result rate for the 205.

TABLE 1

Maximum computation rates in MFLOPS *for Cyber 203/205.*

	203 (2 pipe)		205 (2 pipe)	
	64-bit	32-bit	64-bit	32-bit
+	50	100	100	200
×	25	50	100	200
/	12.5	25	100	200

The maximum result rates given above are not achievable because every vector operation has associated with it a delay incurred after the instruction is issued for execution and before the first result emerges from the pipeline. An approximate timing formula for vector instructions for the Cyber 200 has the form

$$(2.1) \qquad\qquad T = S + \alpha n$$

where S is the overhead, frequently called the start-up time, α is the time per result in clock periods, n is the length of the vector, and T is measured in clock periods. For the Cyber 203, the clock period is 40ns, α is $\frac{1}{2}$, 1 and 2 for addition, multiplication and division, respectively, in 64-bit arithmetic and S ranges from 70 clock periods for addition to 165 for division. On the 205, the clock period is 20ns, while $\alpha = \frac{1}{2}$ for 64-bit arithmetic and $S \doteq 50$ for all the arithmetic operations. The effect of the start-up time is to degrade seriously the performance when n is small. This is illustrated in Table 2 for the 205 for a particular case.

TABLE 2

Performance for 2-pipeline Cyber 205 in 64-bit arithmetic.

n	T(clocks)	T/n	MFLOPS
10	55	5.5	9
100	100	1	50
1000	550	.55	91
10000	5050	.505	99
∞	–	.5	100

As Table 2 shows, for short vectors ($n = 10$) performance is less than 10 percent of the maximum rate and vectors of length almost 1000 are required to achieve 90 percent of the maximum rate. The computation rates can be further degraded by the fact that a vector on the Cyber 200 is a set of contiguously addressable storage locations in memory, and if the data for a vector operation is not already stored in such fashion, it must first be rearranged. Although there are hardware instructions (gather, scatter, merge, compress) to effect this data transfer, they add further overhead to the computation.

Hockney and Jesshope [1981] have introduced the useful concept of the half-performance length, $n_{1/2}$, which is defined as the vector length required to achieve one-half the maximum performance; in the example of Table 2, $n_{1/2} = 100$. They use this parameter together with the maximum performance rate to characterize a number of vector and array computers; see also Hockney [1983a, b]. We also mention that it has been noted by several people that "Amdahl's law", first suggested in Amdahl [1967] and a special case of Ware's law discussed in §1, is particularly relevant in parallel computing. Briefly, if a computation contains x scalar operations and y operations that can be done by vector instructions, then the computation can be accelerated by no more than a factor of $(x+y)/x$, even if the vector operations are infinitely fast. For example, if there are 50 percent scalar operations no more than a factor of 2 improvement over scalar code can be achieved.

Cray. The Cray computers are similar in many ways to the Cyber 200 but have fundamental differences. Memory size ranges from 500,000 to 4 million 64-bit words on the Cray-1 and 1S and up to 8 million words on a 4 processor X-MP, but there is no provision for 32-bit arithmetic. Again, there are hardware vector instructions but these utilize separate pipelined arithmetic (functional) units for addition, multiplication, and reciprocation rather than reconfigurable units as on the Cyber 200. The clock period is 12.5ns on the Cray-1 and 1S and 9.5ns on the Cray X-MP, as compared with 20ns on the Cyber 205. The X-MP series allows a configuration of 1, 2 or 4 processors. The most basic functional difference between the processors, however, is that the Cray vector instructions obtain their operands only from eight vector registers, of 64 words each. A schematic of a Cray processor is given in Fig. 2.3.

For data in the vector registers, the vector instructions on the Cray again obey approximately the timing formula (2.1) but now the start-up times are $O(10)$ clock periods, considerably lower than the Cyber 200, while α is, again, $O(1)$. The effect of the small start-up time is evident in Table 3.

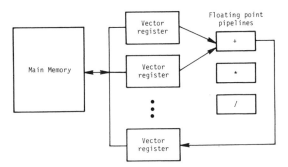

FIG 2.3. *Cray processor schematic.*

TABLE 3

Performance for Cray-1.

n	T(clocks)	T/n	MFLOPS
10	250	25	40
20	375	18.75	53
64	925	14.5	69

Discounting the relatively low start-up time, the maximum result time for each of the three arithmetic operations is 80 MFLOPS on the Cray-1 so that $n_{1/2} = 10$, a significant reduction from the Cyber 205 figure of 100. In addition, all functional units can be operated in parallel so that, theoretically, the maximum speed on the Cray-1 is 160 MFLOPS for addition and multiplication running concurrently. On the X-MP, this figure increases to 210 MFLOPS per processor because of the faster cycle time and would be 840 MFLOPS on a full 4 processor X-MP system.

Although the vector operations on the Cray have relatively low start-up times, this is balanced by the time needed to load the vector registers from memory. For example, for an addition, two vector registers must be loaded and the result stored back in memory from a vector register. Although arithmetic operations can be overlapped to some extent with the memory operations, there is only one path from memory to the registers on the Cray-1 and 1S, and since memory load and store operations require one clock period per word (after a short start-up), only a load or a store can be done concurrently with the arithmetic operations. This problem is alleviated to a large extent on the Cray X-MP series, which has three paths between memory and the registers on each processor, allowing two loads and a store to be done concurrently. In any event, and especially on the Cray-1, one attempts to retain data in the registers as long as possible before referencing memory. One says that *vector speeds* are being obtained if vector hardware instructions are being utilized but that sufficient memory references are required to hold the operation rate to nominally less than 50 MFLOPS on the Cray-1, and that *super-vector speeds* are obtained if information can be held in the registers long enough to obtain rates in the 50-150 MFLOP range. These figures would be scaled up appropriately for the X-MP. The attainment of super-vector speeds is enhanced by the capability of *chaining* which is the passing of results from one arithmetic unit directly to another as illustrated in Fig. 2.4.

Another major advantage that the Cray X-MP offers over the Cray-1 is that of multiple CPU's. The CPU's function off of a shared central memory with a set of shared data and synchronization registers for communication. Some early benchmarks

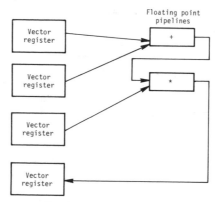

FIG. 2.4. *Chaining on the Cray.*

indicate speedups of from 1.5 to 1.9 for two processors (Chen [1984], and Chen, et al. [1984]).

To summarize the main functional differences between the Cray and Cyber 200, one attempts to organize a computation on the Cyber 200 to utilize vector lengths as long as possible while on the Cray one attempts to organize the computation so as to minimize references to storage and utilize as much as possible information currently in the vector registers. However, any realistic computation will require deviation from these ideals and will also require a certain amount of scalar processing. Several benchmarking studies have been published (e.g. Rudinski and Pieper [1979], Nolen, et al. [1979], Gentzsch [1983], [1984a]) which gave valuable performance data for certain classes of problems. See also Ginsburg [1982].

Japan. It should be noted that Fujitsu, Hitachi and Nippon Electric have developed supercomputers whose performance would appear to be comparable to the Cyber and Cray machines (see, for example, Riganati and Schneck [1984]). The previous discussion on the Cray is appropriate for these machines for they are architecturally similar to it; in particular, they employ vector registers in much the same way as the Cray does (see, for example, Miura and Uchida [1984]). Preliminary benchmark results including a comparison to the Cray are given in Mendez [1984] and Worlton [1984].

Parallel computers. We turn now to computer organizations consisting of a potentially large number of processing elements. These computer architectures fall into two classes as defined by Flynn [1966]. In Single Instruction Multiple Data (SIMD) systems, each processor executes the same instruction (or no instruction) at the same time but on different data. In Multiple Instruction Multiple Data (MIMD) systems, the instructions may differ across the processors, which need not operate synchronously. A much more detailed taxonomy has been given by Schwartz [1983] based on fifty-five designs, and excellent reviews of various architectural approaches are given by Haynes, et al. [1982], Siewiorek [1983] and Zakharov [1984].

Illiac. In the late 1950's, interest in designing a parallel array computer began to grow. Designs such as the Holland Machine (Holland [1959]) and von Neumann's Cellular Automata (von Neumann [1966], first published in 1952), consisting of arrays of processing cells operating in MIMD mode and communicating with their four nearest neighbors, were proposed. In the same period, Unger [1958] had proposed a parallel array of bit-serial processors using the four nearest neighbor communication strategy: the machine was intended for pattern recognition and suggested the architecture of later machines such as the SOLOMON and the DAP. In the early 1960's,

Westinghouse Electric Corp. constructed prototypes of a parallel computer (SOLOMON) designed by Slotnick, et al. [1962]. The design was modified and refined into the Illiac IV at the University of Illinois (Barnes, et al. [1968] and Bouknight, et al. [1972]) and constructed by the Burroughs Corporation.

The Illiac IV consisted of 64 fast processors (about 1 MFLOP each), with memories of 2048 64-bit words connected in an 8×8 array as illustrated in Fig. 2.5. The individual processors were controlled by a separate control unit and all processors did the same instruction (or nothing) at a given time. Hence, the machine was of SIMD type and could be visualized as carrying out vector instructions on vectors of length 64 or shorter. In many ways, algorithm considerations were very similar for the Illiac IV and the Cray and Cyber 200 machines. The Illiac IV was the first parallel array computer to become operational for the benefit of a large, diverse user community when it was installed at the NASA Ames Research Center in the early 1970's. (It was removed from service in 1981.) Although a large number of parallel computers have been, and continue to be, developed, probably most of the computational experience with such computers has been gained on the Illiac IV (see, for example, Feierbach and Stevenson [1979]).

At the same time that the Illiac IV was becoming operational, advances in microprocessors led to a variety of speculations on connecting tens of thousands, or even hundreds of thousands, of such processors together. A major consideration is how these processors are to communicate. The design of the Illiac IV, in which each processor is connected to its four nearest neighbors in the north, south, east, and west directions with wrap-around connections at the edges (Fig. 2.5), is suitable for the simplest discretizations of simple partial differential equations but becomes less suitable for more complicated situations and more sophisticated algorithms. Although the Illiac IV was capable of performing in the 50 MFLOP range, this rate was difficult to sustain because of the relatively small memories and the limitations of the processor interconnection scheme.

Communication. The importance of communication among processors has led to extensive research on interconnection methods. Fundamental work was done by Clos [1953] and by Benes [1962], [1965] for telephone networks, and surveys of more recent research may be found in Anderson and Jensen [1975], Sullivan and Bashkow [1977],

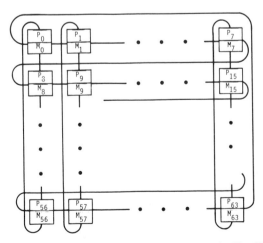

FIG 2.5. *Lattice interconnection as exemplified by the Illiac IV.*

Siegel [1979], Feng [1981], Haynes, et al. [1982] and Broomell and Heath [1983]. It is now clear that the interconnection scheme is probably the most critical issue in the design of parallel systems because it determines how data, and possibly instructions, are made available to the appropriate processing elements. For many algorithms, the total time required to move data to the appropriate processors is as large or larger than the time required for the completion of the computation (see, for example, Gentleman [1978]).

Ideally, every processor would have a dedicated connection to every memory. Although this would allow access in unit time independent of the number of processors, it is impractical for systems with a large number of processors for two reasons. In the first place, the complexity of the interconnection scheme for n processors increases as n^2. Furthermore, since each processor must support n communication lines, if a processor is a single chip or even a few, the number of pins required to provide the communication connections will exceed what present technology can provide, even for moderate size n. On the other hand, an inadequate interconnection scheme limits the performance of the system and thereby reduces the class of problems which can be solved in reasonable time; this is the trade-off facing the designer of a parallel machine.

In practice, three fundamentally different interconnection schemes have been used and, in turn, we will use these to introduce a classification of some simple types of parallel arrays. More complex systems can usually be viewed as a combination of these three types. We also note that, in principle, each of these interconnection schemes could be used to implement a global shared memory.

Lattice. P processors, each with local memory, arranged into some form of regular lattice. Each processor is permanently connected to a small subset of the others, usually its neighbors in the lattice (Fig. 2.5).

Bus. P processors, each with local memory, connected to a bus structure allowing communication among the processors (Fig. 2.6).

Switch. P processors, and M memories connected by an electronic switch so that every processor has access to some, possibly all, of the memories (Fig. 2.7).

Lattice arrays. The classical lattice array is the Illiac IV. Other lattice computers include the Distributed Array Processor (DAP) (Flanders, et al. [1977] and Parkinson [1982]), constructed by International Computers Limited, the Massively Parallel Processor (MPP) at NASA-Goddard (Batcher [1979], [1980]), and the systolic arrays proposed by H. T. Kung and his collaborators (Kung and Leiserson [1979], and Kung [1979], [1980], [1982], [1984]). The DAP is an array of single bit processors, each connected to their four newest neighbors, and with additional row and column data paths. A 64×64 array performing multiplication of two 64×64 matrices using software to effect 32-bit arithmetic provides a computation rate of 18 MFLOPS (Reddaway [1979]). The bit orientation, which permits parallelism at a very low level, and the row and column connections should alleviate some of the communication difficulties of the Illiac IV. The MPP, constructed by Goodyear Aerospace Corp., is also an array of single bit processors, 16,000 of them operating in SIMD mode. It was designed primarily for satellite data reduction but is capable of substantial floating point computation rates. For example, for 32-bit operands, addition may be done at a rate of 430 MFLOPS while the rate for multiplication is 216 MFLOPS (Batcher [1979]). Gallopoulos [1984] discusses performance on several fluid dynamics applications.

Systolic arrays consist of very simple processors capable of performing a single operation such as $ab + c$. They are designed to perform specific computations such as matrix multiplication or LU factorization. This specificity makes it possible to use a

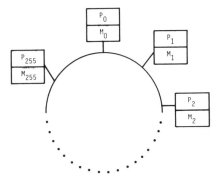

FIG. 2.6. *Bus interconnection as exemplified by* ZMOB.

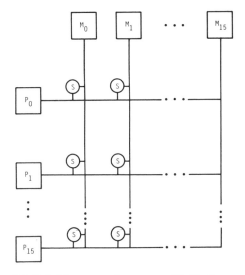

FIG. 2.7. *Switch interconnection as exemplified by* C.mmp.

simple interconnection pattern and move the data continuously through the array. Thus one could view the device as a large pipeline with each processor accepting data, performing a simple operation and passing the operands and/or the result on to the next processing element. The difference between this and a usual pipeline is that each processor performs precisely the same simple function rather than different subfunctions. A significant number of systolic algorithms have been developed; see, for example, Bojanczyk, et al. [1984], Brent and Luk [1983a, b], Heller and Ipsen [1983], Ipsen [1984], Kung [1980], [1984], Kung and Leiserson [1979], Melhem [1983a, b], and Schreiber [1984].

Another lattice that has received considerable attention is the tree structure. For example, Magó [1979], [1980] has proposed a design for directly executing the functional programming languages of Backus [1978] based on a binary tree in which a typical processor is connected to one processor above it and two processors below. Such a tree is said to have a "fan-out" of two; larger fan-outs have been discussed, but there is a potential for communication bottlenecks as the fan-out increases. This can be particularly troublesome at the root of the tree if a large amount of global communication is required. One way to lessen the demand on the root is to introduce horizontal communication links among the processors on any given level of the tree. Shaw [1984] has also proposed a tree structure for the NON-VON machine, which is intended primarily for nonnumerical processing.

It is also possible to consider, at least conceptually, multi-dimensional lattices. An example of such a structure is the binary k-cube for connecting $n = 2^k$ processors (see, for example, Bhuyan and Agrawal [1984]). If the processors are viewed as the corners of a cube in k dimensions, then the connections are the edges of the cube. For $k = 3$ this reduces to a simple cube with each processor connected to three others. In general, if each processor is given a unique label from the integers zero through $n - 1$, then processor i is connected to processor j if and only if the binary representations of i and j differ in a single bit. The strength, and potential weakness, of this strategy is that the number of processors connected to a given processor is k; thus there is a rich inter-connection structure but at some point the requirement for k wires would introduce a fabrication difficulty. A 64-processor machine based on the 6-cube and known as the Cosmic Cube is operational at the California Institute of Technology (see, e.g., Seitz [1982], [1984]). The processors utilize the Intel 8086/8087 chip family and have 128K bytes of memory. In addition, Intel Corp. has announced a commercial version of the machine using the 80286/80287 chip family.

Bus arrays. Examples of bus arrays include Cm* at Carnegie–Mellon University (Swan, et al. [1977], Jones and Gehringer [1980]), ZMOB at the University of Maryland (Rieger [1981]), and Pringle at the University of Washington and Purdue University (Kapauan, et al. [1984]). Cm* is a research system consisting of approximately 50 Digital Equipment Corporation LSI-11's configured in clusters, with the clusters connected by a system of buses. The processors share a single virtual address space and the key to performance lies in the memory references. For example, if the time to service a local memory reference is one unit, then Raskin [1978] reports that a reference to a different memory, but one within the same cluster, requires a little more than three units while a reference to a different cluster requires more than seven units, assuming no contention. Further performance data based on some applications programs, including the iterative solution of a discretized Laplace's equation and a problem in computational chemistry, are given in Raskin [1978] and Hibbard and Ostlund [1980], respectively. Additional applications are treated in Ostlund, et al. [1982], and general programming considerations are discussed in Jones, et al. [1978]. The machine was also used to simulate electrical power systems (Dugan, et al. [1979] and Durham, et al. [1979]).

ZMOB is an array of up to 256 Z-80 microprocessors configured in a ring as depicted in Fig. 2.6. The performance of the bus relative to that of the processors is so great that there are not the usual delays in communication characteristic of bus arrays. Because of the high speed bus, a processor can obtain data from any memory in approximately the same time but, unfortunately, this is not a characteristic that could be maintained if the array were scaled up to a larger number of more powerful processors.

The Pringle system was designed and built to serve as a test bed to emulate the CHiP architecture (Snyder [1982]) as well as others. The system consists of 64 processing elements based on 8-bit Intel processors with a floating point coprocessor (Field, et al. [1983]). The processing elements, with a modest amount of local memory, are connected via separate input and output buses. The two buses are connected via a message routing processor or "switch" which establishes communication patterns that allow the Pringle to emulate a variety of communication networks. Some preliminary performance data is given in Kapauan, et al. [1984] for summing a sequence of numbers using an algorithm based on recursive doubling.

Switch arrays. The now classic example of a switch array is C.mmp, a research machine developed at Carnegie–Mellon University in the early 1970's (Wulf and Bell [1972] and Wulf and Harbison [1978]). The system consisted of up to sixteen Digital Equipment Corporation PDP minicomputers connected to sixteen memory modules via a 16×16 crosspoint switch, as depicted in Fig. 2.7. There is not a great deal of data on the performance of C.mmp on scientific applications; however, one study by Oleinick and Fuller [1978] provides insight into the importance of synchronization on performance. In a parallel version of the bisection method for finding the root of a monotonically increasing function, after all processors have evaluated the function they must halt and await the decision of which subinterval to use next. Several synchronization techniques were investigated and it was found that their time of execution varied by a factor of 15 with the more sophisticated techniques requiring over 30 milliseconds. This obviously adds significant overhead to the algorithm for all but the most complicated functions. Synchronization techniques are a major area of concern in the design and use of parallel arrays.

The crosspoint switch is also the basis for the communication mechanism for the S-1 array under development at Lawrence Livermore National Laboratory (Farmwald [1984]). This machine is intended to support up to sixteen processors of approximately Cray-1 performance connected to a shared memory consisting of about 10^9 bytes per processor.

The full crosspoint switch for connecting n processors with n memories contains n^2 switches, which is not feasible for large n. This has led designers to consider simpler switches consisting of $O(n \log n)$ subswitches. An introduction to this area is contained in Haynes, et al. [1982]. An example of an $n \log n$ switch (and there are many) is the Banyan switch (Goke and Lipovksi [1973]), which is the basis for the Texas Reconfigurable Array Computer (TRAC) under development at the University of Texas (Sejnowski, et al. [1980] and Browne [1984b]). Some projected performance data for the TRAC, based on odd-even reduction algorithms for block tridiagonal systems (Heller [1976]), is given by Kapur and Browne [1981], [1984].

Another computer utilizing a Banyan type switch is the Heterogeneous Element Processor (HEP) manufactured by Denelcor, Inc. (Smith [1978] and H. Jordan [1984]). It consists of up to sixteen processors with the switch providing access to a data memory. In a HEP processor two queues of processes are maintained. One of these controls program memory, register memory and the functional units while the other controls data memory. The mode of operation is as follows. If the operands for an instruction are contained in the register memory, the information is dispatched to one of several pipelined functional units where the operation is completed; otherwise the process enters the second queue which provides information to the switch so that the necessary link between processor and data memory can be established. After the memory access is complete, the process returns to the first queue, and when its turn for service occurs, it will execute since the data is available. The time required to complete an instruction is 800 ns, but a new instruction may be issued every 100 ns. Thus, if the processors can be kept fully utilized, on a sixteen processor machine a 160 MFLOP rate is theoretically possible. Some preliminary information on utilizing the HEP for solving linear systems is given by Lord, et al. [1980], [1983], H. Jordan [1983], [1984], Dongarra and Hiromoto [1984] and by H. Jordan [1981], who concentrates on the sparse matrix package from AERE, Harwell, England (Duff [1977]). Moore, et al. [1985] discuss several hydrodynamic applications for which efficiencies very close to unity are obtained on a single processor HEP. Operational results for a four processor HEP at the Army Ballistic Research Laboratory are given in Patel and Jordan [1985], where an iterative method for a problem in fluid mechanics is discussed. Some preliminary performance data for the HEP and several other MIMD systems may be found in Buzbee [1984b].

Other connection schemes. We have indicated many examples of parallel array computers covering the three major classifications. Because of the importance of communication and the limitations of the various strategies, we may expect to see computers which utilize combinations of the techniques described above. One such machine is the Finite Element Machine (H. Jordan [1978a], Storaasli, et al. [1982] and Adams and Voigt [1984b]) at the NASA Langley Research Center. This lattice array was designed for 36 16-bit microprocessors configured in a planar array with each processor connected to its eight nearest neighbors as shown in Fig. 2.8. It is also a bus array because the nearest neighbor connections are augmented by a relatively high performance bus which services every processor of the array. Some preliminary performance data are available in Adams [1982] and Adams and Crockett [1984]. A rather similar machine, PACS, is being developed in Japan (Hoshino, Kawai, et al. [1983] and Hoshino, Shirakawa, et al. [1983]).

Another system which combines communication strategies is MIDAS, a prototype of which is operational at the Lawrence Berkeley Laboratories (Maples, et al. [1983]). The array consists of clusters of processors configured as a tree, with each cluster containing up to eight processors interconnected by a full crosspoint switch. A discussion of programming considerations and some results for a Monte Carlo simulation are given in Logan, et al. [1984]; additional results are reported in Maples, et al. [1984].

Another approach to the communication problem is the configurable highly parallel (CHiP) computer (Snyder [1982]). An interesting feature of this project is that the communication network is programmable and reconfigurable. Thus, for an application involving the direct solution of linear systems, it can be made to function as an appropriate systolic device, while for another application involving the iterative solution of a discretized differential equation, it can function as a lattice array. Reconfigurability offers additional benefits for adapting to varying problem sizes and for providing fault tolerance. An example of this flexibility is given in Gannon and Panetta [1985] which discusses implementing SIMPLE, a benchmark hydrodynamics code, on the CHiP.

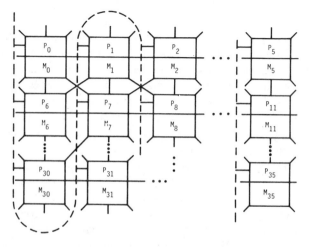

FIG. 2.8. *Finite Element Machine.*

There are also several other efforts whose impact will have to await further development. More exotic switching networks such as the shuffle exchange, cube-connected-cycles and omega networks have been studied (see, for example, Haynes, et al. [1982]). The Ultracomputer project at New York University is building, in cooperation with IBM, a large array based on the shuffle exchange (Gottlieb and Schwartz [1982], Gottlieb, Grishman, et al. [1983] and Gottlieb [1984]). The nodes of the shuffle exchange network possess rudimentary processing capability which is used to help alleviate memory contention. The Cedar project (Gajski, et al. [1983], [1984]) at the University of Illinois makes use of the omega network to connect clusters of processors to a large shared memory. Wagner [1983] has proposed the Boolean Vector Machine (BVM) as a large array of single-bit processors with very small memories operating in SIMD mode using bit serial arithmetic. The processors are interconnected via a cube-connected-cycles network (Preparata and Vuillemin [1981]) which links each processor to three others. An algorithm for solving sparse linear systems on the BVM is analyzed in Wagner [1984].

Another interesting idea is the dataflow computer, which has been the subject of over 10 years of research by J. Dennis of MIT and his colleagues as well as others (see e.g. Dennis [1980], [1984b] and Agerwala and Arvind [1982] and the references therein for a general overview). Two systems based on this concept are the Manchester Data Flow Machine (Watson and Gurd [1982]) which has been operational since 1981, and the SIGMA-1 (Hiraki, et al. [1984] and Shimada, et al. [1984]) which is under construction at the Electrotechnical Laboratory in Japan. Gurd and Watson [1982] report very promising results for a variety of problems run on the Manchester machine. Studies on the effectiveness of data flow computers for applications such as the weather problem and computational fluid dynamics have been done by Dennis and Weng [1977] and Dennis [1982], [1984a], respectively.

It is now clear that new designs from commercial manufacturers will utilize a combination of vector and array concepts for computers that might be characterized as arrays of vector processors. The first of these was the Cray X-MP (see e.g. Chen [1984]), which was introduced as a two processor version and is now available with four processors. The Cray-2 and Cray-3 are also expected to involve multiple processors with the Cray-2 initially offering four. Control Data Corp. has announced the Cyberplus (Ray [1984]) which consists of up to 64 processors each with multiple pipelined functional units. The functional units within a processor may be connected via a crossbar switch to obtain the equivalent of chaining, and the processors themselves are connected via three independent buses. ETA Systems Inc., a spin-off of Control Data Corp., has announced the GF-10 (Johnson [1983]). This system is expected to utilize up to eight pipelined processors similar to the Cyber 205, but faster, operating off of a shared memory with up to 256 million words. The individual processors will also have local memory of approximately four million words.

3. Direct methods for linear equations. We consider in this section direct methods for solving linear algebraic systems of equations

$$(3.1) \qquad\qquad A\mathbf{x} = \mathbf{b}$$

where A is $n \times n$. Our main concern will be when A is banded and usually symmetric positive definite (or at least pivoting is not required). We will treat first elimination (factorization) methods, then methods based on orderings such as nested dissection, and finally special methods for tridiagonal systems and fast Poisson solvers.

Elimination methods. Consider first Gaussian elimination, without pivoting, when A is a full matrix. If we assume that A is stored by columns, as done by Fortran, then the usual row-oriented elimination process is not suitable for vector machines. Rather, we need a column-oriented algorithm as illustrated by the following first step of the elimination process. Let $\mathbf{a}_i$ be the $n-1$ long vector of the last $n-1$ elements of the ith column of A. Then

$$(3.2) \qquad \mathbf{m} = a_{11}^{-1}\mathbf{a}_1, \quad \mathbf{a}_i - a_{1,i}\mathbf{m} \to \mathbf{a}_i, \quad i = 2, \cdots, n$$

completes the first step of the reduction. Note that all operations except one are $n-1$ long scalar-vector multiplies or vector additions.

Following Hockney and Jesshope [1981], we will say that the *degree of parallelism* of an algorithm is the number of operations that can be done concurrently. On vector computers, such as the Cyber 200 and Cray, we will interpret this to mean the vector lengths while on parallel computers it will mean the number of processors that can be operating simultaneously. Clearly the degree of parallelism is $n-1$ for the first stage of the elimination reduction. For the second stage, the vector lengths decrease to $n-2$ and so on down to a vector length of 1 for the last stage. Hence, the degree of parallelism constantly decreases as the reduction proceeds, with an average degree of parallelism of $O(3n/2)$ since there are $n-j$ vector operations of length $n-j$ at the jth stage.

If A is banded, with semi-bandwidth m defined by $m = \max\{|i-j|: a_{ij} \neq 0\}$, then the above algorithm allows constant vector lengths m until the reduction has proceeded to the last $m \times m$ block, at which time the vector lengths again decrease by one at each stage down to a length of 1. Thus, this algorithm leads to a low degree of parallelism for small m and is totally inappropriate for tridiagonal matrices ($m=1$), for which special methods will be discussed later in this section.

While the above form of Gaussian elimination is an appropriate starting point for a parallel algorithm, the architectural details of a particular machine may necessitate changes, perhaps drastic, to achieve a truly efficient algorithm. Several early papers (e.g. Lambiotte [1975], Knight, et al. [1975], Calahan, et al. [1976], George, et al. [1978b], Fong and Jordan [1977]) considered in detail the implementation of Gaussian elimination and the Choleski decomposition $A = LL^T$ on the CDC STAR-100, TI-ASC, and Cray-1. The variations on the basic algorithms because of the machine differences are summarized in Voigt [1977].

An important aspect of the analysis in some of the above papers is the derivation of precise timing formulas which show the effect of the start-up times for vector operations. For example, George, et al. [1978b] gave the following formula, which omits scalar arithmetic times, for the Choleski decomposition of a banded matrix, taking advantage of symmetry in the storage, on the STAR-100.

$$(3.3) \qquad T = 0.75nm^2 + 232nm + \text{low order terms}.$$

This timing formula is in units of machine cycles. The leading term reflects the arithmetic operation count and the result rate for addition and multiplication while the second term shows the effect of the vector operation start-up times which contribute most of the large coefficient of the nm term. As an example of the effect of machine architecture, Voigt [1977] showed that by modifying the Choleski algorithm to take advantage of some features of the TI-ASC, the timing formula on that machine became

$$T = nm^2 + \frac{19}{16}nm + 485n + \text{low order terms}$$

which gave a dramatic decrease in the coefficient of the nm term. Timing formulas analogous to (3.3) can be developed for the Cyber 205 and show a similar, but smaller, effect of start-up time in the second term.

On the Cray-1, one is much less concerned with the start-up times; instead the basic Choleski or elimination algorithms must be revised to keep data in the vector registers as long as possible. This is accomplished by completely modifying the $(k+1)$st column of the matrix during the kth step of the factorization, leaving all other columns unchanged. The details may be found in Voigt [1977] for the Choleski algorithm and in Dongarra, Gustavson and Karp [1984] for Gaussian elimination. The latter paper gives an interesting detailed analysis of six different forms of the basic algorithm which differ only in how the data is accessed.

The above discussions concern only the factorization phase of the overall algorithm and it still remains to carry out the forward and backward substitutions, i.e. to solve lower and upper triangular systems. Perhaps the simplest and most natural approach to this, called the *column sweep* algorithm in Kuck [1976], is as follows for the upper triangular system $Ux = b$. First, x_n is computed from the last equation and its value is inserted into each of the remaining equations so as to modify the right-hand side, and, clearly, the $n-1$ equations can be processed in parallel. The original system is now reduced to an $n-1 \times n-1$ system and the process is repeated. The degree of parallelism is the bandwidth m until the system has been reduced to $m \times m$ and then the degree of parallelism is reduced by one at each stage. We will consider other algorithms for triangular systems later.

One way to circumvent, in a sense, the back substitution phase is by the Gauss–Jordan algorithm, which is not often used on serial computers since its operation count of $O(n^3/2)$ to solve a linear system has a larger constant than the $O(n^3/3)$ of Gaussian elimination. However, it is relatively more attractive for parallel computing since the back substitution is effectively combined with the triangular reduction in such a way that a degree of parallelism of order n is maintained throughout the computation. The implementation of the Gauss–Jordan algorithm on arrays of processors has been discussed by Kant and Kimura [1978] and Kimura [1979]; see also Parkinson [1984] for a banded system. Unfortunately, the algorithm fills in the upper triangle and so is not attractive for a banded system.

In principle the factorization methods discussed above may be implemented on parallel arrays and a nice introduction may be found in Heller [1978]. For example, the vector operations in expression (3.2) could be given to the ith processor, $i = 2, \cdots, n$, or if fewer processors are available, groups of columns could be assigned to processors. It should also be noted that the more usual row-oriented elimination could be implemented in a similar fashion. But these algorithms have at least three drawbacks. First, as was pointed out above, the degree of parallelism decreases at each stage of the elimination, eventually leaving processors unused. Second, the algorithms require significant communication because the pivot column (row) must be made available to all other processors (see e.g. Adams and Voigt [1984a]). Third, when the problem does not match the array size, a very difficult scheduling problem may arise (see, e.g., Srinivas [1983]). For banded matrices the processor utilization problem is not as severe since it is not a factor except in the final stages.

A detailed analysis of the computational complexity of factorization algorithms may be found in Kumar and Kowalik [1984]. Algorithms for the Denelcor HEP are given in Dongarra and Hiromoto [1984] and the banded case is discussed in Dongarra and Sameh [1984]. Computational results are reported by Leuze [1984b] for the Finite

Element Machine, and an interesting aspect of this study is the influence that different organizations of the rows of the matrix have on the performance of the algorithm due to different communication requirements. Leuze [1984a] and Leuze and Saxton [1983] have also noted that minimizing the bandwidth does not always lead to the best parallel factorization time for a banded matrix. They suggest other orderings of the matrix which appear to improve on the degree of parallelism. Huang and Wing [1979] present a heuristic for reordering a matrix specifically to increase the degree of parallelism. They also discuss an implementation on a hypothetical parallel system designed to take advantage of the heuristic.

Algorithms based on a block partitioning of A are natural to consider on arrays of processors. Lawrie and Sameh [1983], [1984] (see also Sameh [1983] and Dongarra and Sameh [1984]) give a block elimination algorithm for symmetric positive definite banded systems which generalizes one of Sameh and Kuck [1978] for tridiagonal systems. The coefficient matrix A is partitioned into the block tridiagonal form

$$A = \begin{vmatrix} A_1 & B_1 & & \\ B_1^T & A_2 & \ddots & \\ & \ddots & \ddots & B_{p-1} \\ & & B_{p-1}^T & A_p \end{vmatrix}$$

where each B_i is strictly lower triangular and p is the number of processors. For simplicity, assume that each A_i is $q \times q$ so that $n = pq$. The factorizations $A_i = L_i D_i L_i^T$ are then carried out in parallel, one per processor. Using these factorizations, the systems $A_i V_i = B_i$, $A_{i+1} U_{i+1} = B_i^T$, $i = 1, \cdots, p-1$, are solved, utilizing the zero structure of the B_i. These solutions are done in parallel, one pair per processor. The matrix A has now been reduced to

$$\hat{A} = \begin{vmatrix} I & V_1 & & \\ U_2 & I & V_2 & \\ & \ddots & \ddots & \ddots & V_{p-1} \\ & & U_p & I \end{vmatrix}$$

and, provided that $2pm \leq n$, where m is the semi-bandwidth of the system, there is an uncoupling of $2m(p-1)$ equations in the corresponding system, namely the equations $jq - m + k, j = 1, \cdots, p-1, k = 1, \cdots, 2m$. Once this smaller system is solved, the remaining unknowns can be evaluated by substitution. Note that the larger $2mp$, the larger the size of the uncoupled system, which is undesirable. A reasonable balancing of work would have all systems roughly the same size. Since the A_i are $n/p \times n/p$ this would imply that $2mp \doteq n/p$ or $2mp^2 \doteq n$ which, of course, also implies that $2mp < n$.

Other block algorithms have been proposed by Hwang and Cheng [1980] and Halada [1980], [1981]. The former authors, motivated by VLSI design, propose a block Gaussian elimination scheme in which four basic chips handle LU decomposition without interchanges, matrix multiplication, matrix-vector multiplication, and inversion of triangular matrices, respectively. Halada presents an algorithm for banded linear systems with $n > 3m$ based on the partitioning of the system as

$$\begin{bmatrix} A_{11} & A_{12} \\ 0 & A_{22} \end{bmatrix} \begin{bmatrix} x_1 \\ x_2 \end{bmatrix} = \begin{bmatrix} b_1 \\ b_2 \end{bmatrix}$$

where A_{12} is $n - m \times n - m$, triangular, and assumed nonsingular. The key step in the algorithm solves an auxiliary system with the coefficient matrix

$$\begin{bmatrix} A_{12} & 0 \\ A_{22} & I \end{bmatrix}.$$

Unfortunately, without further (and unreasonable) assumptions on A_{12}, the algorithm is numerically unstable.

The above discussions are predicated primarily on the assumption that A is symmetric positive definite or, in any case, that no interchanges are required to maintain numerical stability. The incorporation of an interchange strategy into Gaussian elimination causes varying degrees of difficulty on parallel architectures. Partly to alleviate these difficulties, Sameh [1981] (see also Sorensen [1985] for further analysis) introduced a different pivoting strategy in which only two elements at a time are compared. This ensures that the multipliers in the elimination process are bounded by 1, but requires an annihilation pattern different from the usual one for Gaussian elimination. (This annihilation pattern is identical to the one used for the parallel Givens algorithm of Sameh and Kuck, to be discussed next.)

Givens reduction. The difficulties with implementing interchange strategies on parallel architectures suggest that orthogonal reductions to triangular form may have advantages. It was observed by Gentleman [1975] that the orthogonal reduction to triangular form by Givens or Householder transformations has a certain natural parallelism, and an algorithm for the Givens reduction was given in detail by Sameh and Kuck [1978], who also show that the use of Givens transformations is slightly more efficient in a parallel environment than Householder transformations. Recall that the Givens reduction to triangular form can be written as

$$Q_r \cdots Q_1 A = U$$

where $r = n(n-1)/2$ and each Q_i is a plane rotation matrix whose multiplicative effect is to zero one element in the lower traingular part of A. The Sameh–Kuck algorithm groups these rotations in such a way as to achieve a degree of parallelism essentially the same as Gaussian elimination. An illustration of the grouping is given in Fig. 3.1 for an 8×8 matrix in which only the subdiagonal elements are shown. In this figure, the number indicates the stage at which that element is annihilated. Gannon [1980] develops an implementation of the Sameh–Kuck Givens algorithm for a mesh-connected array of processors such as the Finite Element Machine. The implementation is such

```
7
6   8
5   7   9
4   6   8   10
3   5   7   9   11
2   4   6   8   10  12
1   3   5   7   9   11  13
```

FIG. 3.1. *Sameh–Kuck Givens annihilation pattern*.

that the data moves through the array so as to give a pipelining or systolic effect. The back solve is carried out in an analogous way.

Lord, et al. [1980], [1983] (see also Kowalik, Kumar and Kamgnia [1984]) also discuss Givens transformations for full systems, motivated by multiprocessor systems and the Denelcor HEP in particular. As opposed to the annihilation pattern of Sameh and Kuck [1978], which is predicated on using $O(n^2)$ processors, they assume that $p \leq O(n/2)$ and give two possible annihilation patterns as illustrated in Fig. 3.2. The zigzag annihilation pattern is based on using $(n-1)/2$ processors, one for each two subdiagonals, while the column sweep pattern assumes $p \ll n$. Numerical results indicating the effectiveness of the zigzag algorithm on the Denelcor HEP are given in Lord, et al. [1980]. Although not discussed by the authors, note that the zigzag pattern adapts nicely to banded systems; here one would assume that $p = \lceil m/2 \rceil$. Moreover, for banded systems the process is relatively more efficient since in the full case, the higher numbered processors are doing considerably less work. The column sweep pattern also adapts nicely to banded systems and seems to be very efficient. Other parallel orderings for Givens annihilations are considered by Modi and Clarke [1984].

In general, it would seem that the use of Givens transformations could be preferable on some architectures to Gaussian elimination if interchanges are required and not otherwise. For least squares problems, however, orthogonal reduction has other advantages, and Sameh [1982] considered the use of Givens transformations for this problem in the context of a ring network of processors.

Some other methods. Toeplitz matrices (each diagonal is constant) arise in a number of applications. Grcar and Sameh [1981] consider banded Toeplitz matrices and, under various assumptions on the matrix, they give three algorithms. For banded symmetric positive definite matrices, their algorithm requires $O(m \log n)$ time steps using $4n$ processors. See also Bini [1984] for other work on Toeplitz matrices.

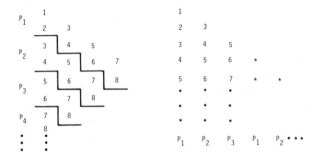

FIG. 3.2. *Givens annihilation patterns and processor assignments*.

An interesting variation of the elimination process has been advanced by D. Evans and several colleagues in a series of papers (Evans and Hatzopoulos [1979], Evans and Hadjidimos [1980], Evans, et al. [1981], Shanehchi and Evans [1981], [1982]) and reviewed in Evans [1982b], [1983]. The basic idea is a factorization of A, called the Quadrant Interlocking Factorization (QIF), which has the structure

$$(3.4) \quad A = WZ = \begin{bmatrix} 1 & 0 & & \cdots & & 0 \\ * & 1 & & & & * \\ & & \ddots & 0 & 0 & \\ \vdots & & * & 1 & * & \vdots \\ & & 0 & 0 & \ddots & \\ * & & & & * & \\ 0 & & \cdots & & 0 & 1 \end{bmatrix} \begin{bmatrix} * & & \cdots & & * \\ 0 & & & & 0 \\ \vdots & & 0 & * & 0 & \vdots \\ 0 & & & & 0 \\ * & & \cdots & & * \end{bmatrix}.$$

Here W has 1's on its main diagonal, Z has nonzeros on its main diagonal and the $*$'s indicate generally nonzero elements. Variations of this factorization have been given that allow a Choleski type decomposition WDW^T and that are appropriate for banded systems.

The decomposition (3.4) is carried out as follows. First $z_{1i} = a_{1i}$, $z_{ni} = a_{ni}$, $i = 1, \cdots, n$, and then the first and last columns of W are obtained from the $n-2$ 2×2 systems

$$(3.5) \quad \begin{aligned} w_{i1} z_{11} + w_{in} z_{n1} &= a_{i1}, \\ w_{i1} z_{1n} + w_{in} z_{nn} &= a_{in}, \end{aligned} \qquad i = 2, \cdots, n-1.$$

The first and last columns of W and Z are now determined and the elements of A are updated by

$$(3.6) \qquad A \rightarrow A - W_1 Z_1^T - W_n Z_n^T$$

where W_1 and W_n are the first and last columns of W, and Z_1^T and Z_n^T the first and last rows of Z. The first stage of the factorization is now complete and the second stage proceeds in the analogous way to determine the remaining elements in the 2nd and $(n-1)$st rows and columns of W and Z and then to update A corresponding to (3.6). Thus the factorization is complete in $O(n/2)$ stages.

At the kth stage, $n - 2k$ 2×2 systems need to be solved to determine the w's at that stage, and these 2×2 systems can be solved in parallel. Also, an $n - 2k \times n - 2k$ submatrix of A needs to be updated and these calculations can also be done in parallel. Hence, the degree of parallelism at the kth stage is $O(n - 2k)$ and the overall average degree of parallelism is $O(n/2)$. To complete the solution of $Ax = b$, we then need to solve the systems $Wy = b$ and $Zx = y$. The solution of the first system can be overlapped with the factorization; as the w's become available during the factorization, the corresponding y's can be computed.

Evans and his coworkers have done various analyses of this and related QIF methods and claim essentially the same numerical stability as Gaussian elimination; in particular, the algorithms are stable if A is symmetric positive definite or diagonally dominant. These QIF methods seem to be potentially attractive alternatives to Gaussian elimination or Choleski factorization for parallel computation but more experience with their numerical stability and efficiency on different parallel architectures is needed.

Maximum parallelism. The methods we have reviewed so far all have a maximum degree of parallelism of $O(n)$ for full systems or $O(m)$ for banded systems. There have been a number of attempts, especially in the earlier literature, to devise methods with a higher degree of parallelism. In general, these papers have been directed at the theoretical question of how fast a system can be solved given an unlimited number of processors, ignoring such practical constraints as communication delays, etc. Several results of this kind are reviewed in detail in Sameh [1977] and Heller [1978] and we give here only a sampling.

It is quite easy to see that, for a full matrix and without pivoting, Gaussian elimination can be carried out in $3(n-1)$ time steps using $(n-1)^2$ processors. Preparata and Sarwate [1978], improving on a result of Csanky [1976], showed that the system can be solved in $O(\log^2 n)$ time steps using no more than $2n^{3.31}/\log^2 n$ processors. The algorithm makes use of the Cayley–Hamilton theorem to compute A^{-1} and is numerically unstable. It is an interesting complexity result but does not yield a practical algorithm.

Triangular systems. Similarly, for triangular systems (which are to be solved in the back substitution phase of elimination or orthogonal reduction algorithms), Sameh and Brent [1977] gave algorithms which could be carried out in $O(\log^2 n)$ steps using no more than $n^3/68 + O(n^2)$ processors for full matrices, and $O((\log m)(\log n))$ steps using no more than $m^2n/2 + O(mn)$ processors if the bandwidth is m. These results improved on previous ones of Chen and Kuck [1975], but the error analysis given, as well as some numerical results, shows that the algorithms may be numerically unstable in certain cases. Chen, et al. [1978] gave another algorithm for banded systems which requires $O(2m^2n/p)$ time steps, where p, the number of processors, is assumed to be at least $2m$. Generally, this algorithm will require more time steps, but uses fewer processors; for example, if $p = 2m$, $O(mn)$ time steps are required. Again, an error analysis performed by the authors showed a potential exponential growth in rounding error, but numerical experiments indicated that these error bounds were probably unrealistically large.

More recently, Montoye and Lawrie [1982] have given an algorithm for full triangular systems on a hypothetical SIMD array of p processors which are connected to p memories with suitable alignment networks. The algorithm uses partitioning of the system and requires $O(n^{2-r})$ time steps with $r = \log p/\log n$; for example it requires $O(n)$ steps with n processors.

Evans and Dunbar [1983] give two algorithms for solving triangular systems called the Wavefront and Delayed Wavefront methods. The former assumes that the number of processors satisfies $2(n-1)/3 \leq p \leq n-1$ while the latter assumes that $p < 2(n-1)/3$. In both cases, optimal performance is achieved for $p = 2(n-1)/3$ and, in this case, $O(2n)$ time steps are required. The algorithms proceed in 3 phases. In the first, processors are assigned to the 2nd through $(p+1)$st rows of the system. The known value x_1 is substituted into row 2, giving x_2, and processor 2 is reassigned to row $p+2$. The process continues in this way until a processor has been assigned to row n. This is the end of the first phase. In the second stage, as soon as processor k becomes available it is reassigned to row n at column $k+1$, processor $k+1$ is assigned to row $n-1$ at column $k+2$, and so on. The x_i are now being worked on in two pieces until the two "wavefronts" come together. At this point, there remains only a triangular system of less than p rows and it is solved by assigning one row per processor. A potential drawback of these methods is the large amount of communication required by reassigning processors.

Although many of the above algorithms for triangular systems are interesting, and may turn out to be useful in practice, it is unlikely that they will give enough speedup over the basic column sweep algorithm to justify their increased complexity. Moreover, the numerical stability of the column sweep algorithm is well understood since it is just the usual serial algorithm.

Nested dissection. The previous discussion has focused on banded systems such as might arise from discretizations of elliptic equations in which the node points are ordered so as to achieve relatively small bandwidths. We now consider other orderings that are known to reduce both the number of arithmetic operations and the storage requirements for factoring the matrix of the resulting system. The first of these is known as one-way dissection and is discussed in detail in George [1972], [1977] and George and Liu [1981]. Referring to Fig. 3.3, the idea is first to divide the grid of $N \times N$ nodes with l horizontal separators. The nodes in the $l+1$ remaining rectangles are numbered toward a separator as indicated by an arrow and then the separators are numbered. For the proper choice of l this ordering has been shown (see George [1972]) to reduce the number of arithmetic operations required for the factorization of the $n \times n$ ($n = N^2$) system from $O(n^2)$ for the natural ordering to $O(n^{7/4})$.

The nested dissection ordering further reduces the operation count to $O(n^{3/2})$ as shown in George [1973], [1977]. The idea here is to divide the grid with both horizontal and vertical separators as shown in Fig. 3.4. Regions 1–4 are again divided using horizontal and vertical separators. Clearly the idea may be applied recursively, and in the case $N = 2^k - 1$, dissection will terminate after $k-1$ steps. In order to obtain the $O(n^{3/2})$ operation count, dissection must be carried to completion; however, as noted in George, et al. [1978a], there are advantages in terms of storage to terminating dissection early.

Nested dissection for vector computers was first discussed by Calahan [1975] in the context of rather general rectangular finite elements, and estimates are given of the number of vector operations required for the factorization and their average lengths, assuming dissection is carried to completion. The appropriate level of dissection becomes an interesting question for a vector computer. We have already seen that for the Cyber 200 it is desirable to work with vectors whose length is as great as possible;

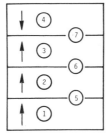

FIG. 3.3. *An $N \times N$ mesh dissected into 4 blocks with the ordering indicated by the circled numbers and the arrows.*

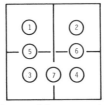

FIG 3.4. *One step of the nested dissection ordering for the $N \times N$ grid.*

however, from Figs. 3.3 and 3.4 it is clear that at least some of the vectors become shorter as dissection continues. This problem is studied in detail in George, et al. [1978b] for vector computers with a range of start-up times covering both the Cray and the Cyber 200. For the Cyber 200 their results indicate that the minimum time for factorization is obtained by stopping nested dissection two levels from completion. Another approach to the problem of vector lengths was suggested by Calahan [1975]. He noted that on the TI-ASC it was possible to execute simple triply nested DO loops as one vector instruction resulting in vector lengths equal to the product of the loop lengths. Applying this idea to nested dissection resulted in an increase in average vector lengths. Unfortunately no vector computer presently available provides this capability; however, the idea is closely related to unrolling DO loops, a technique that has become a powerful way to increase performance (see §1).

Another result discussed in George, et al. [1978b] deals with the general problem of how effectively an algorithm translates into vector operations. Both the one-way and the nested dissection algorithms translate almost entirely into vector operations; however, in spite of a lower operation count, one-way dissection introduces more vector operations than are present in banded algorithms for the natural ordering, resulting in the natural ordering being superior for all but very large n. Fortunately this phenomenon is fairly rare, and as expected, nested dissection can be implemented with fewer vector operations than the usual banded algorithms. This situation is discussed in more detail in Voigt [1977]. In principle, both dissection algorithms would be attractive for the Cray X-MP; however, the limited paths between memory and the vector registers could adversely affect performance on the earlier Crays.

Calahan [1979b] introduces a variant of nested dissection in which the separators are the diagonals indicated by the square points in Fig. 3.5. The dissection may be performed recursively, and Calahan claims that if the nodes are properly ordered, resulting in 4×4 diagonal blocks, the process may be implemented on the Cray-1 with performance in excess of 50 MFLOPS on the 4×4 factorizations. The dense lower right-hand block corresponding to the separators may be factored at rates in excess of 100 MFLOPS. For the processing required by the blocks that represent the connections between the separators and the 4×4 diagonal blocks, Calahan estimates performance in the 30 MFLOP range but this requires introducing new nodes and unknowns in order to achieve regularity of the block structure. This technique adds approximately 25 percent more nodes to the dense lower right block and the overall effect on computation time is not known.

Another variant of nested dissection suggested by Liu [1978] may offer distinct advantages for parallel arrays. Liu suggests making the separators two mesh lines wide rather than one as in the George algorithm shown in Fig. 3.4. This provides more complete independence of the remaining subsets which may lead to better interprocessor communication characteristics. A possible disadvantage is that the submatrices associated with the separators are twice as large. Nevertheless, Liu shows that, in theory, the algorithm will solve an $N \times N$ grid problem in $O(N)$ steps using $O(N^2)$ processors.

FIG. 3.5. *Diagonal variant of nested dissection*.

For parallel arrays, a careful analysis of nested dissection has been given by Gannon [1980]. He considers an MIMD array with nearest neighbor connections and assumes a processor for each node in the discretization. The algorithm uses a pipelined version of Givens rotations as a building block. Utilizing an $N \times N$ array for an $N \times N$ grid, Gannon shows that nested dissection will run in $C(N + r \log N)$ time for r right-hand sides. The constant C is fairly large and may result in the algorithm not being competitive with other methods for a single right-hand side. Communication time is included, and he shows that contention for data in common regions such as the bisectors can be avoided. The MIMD capability is essential because different processors execute different code sequences. Allocation of one grid point per processor does mean that some processors would be idle during the algorithm. Using more points per processor could increase processor utilization but it might also increase communication time.

General sparse matrices. We now turn to general sparse matrices. The methods discussed above do not explicitly deal with the sparsity structure of the system (3.1). For banded matrices this is not normally necessary because the matrix fills out to the band during the factorization. However, there are applications such as load in electrical power networks which produce very sparse matrices with little exploitable structure, and treating these as dense systems incurs an intolerable overhead. The importance of such systems was recognized by Control Data Corp. in the development of the STAR-100 and sparse arithmetic instructions were implemented; these remain available on the Cyber 200. The idea is to store as vectors only the nonzero values, together with a bit vector which indicates the location of the nonzero elements. There seems to be little use of the instructions, however, because their performance is not much better than the standard arithmetic instructions unless the vectors are extremely sparse and the non-zeros occur in clusters. In addition, the storage requirements of the bit vector are much greater than those of modern sparse matrix methods. For example, since a word on the Cyber 200 contains 64 bits, the storage of an $n \times n$ sparse matrix requires $n^2/64$ words plus the nonzeros even though the matrix may be less than 1 percent dense. For large matrices this is simply too large an overhead.

The storage requirement is potentially reduced in a sparse arithmetic processor proposed by Gao and Wang [1983]. In their scheme, an integer vector denoting the locations of the nonzero elements of the data vector is carried with the data vector. Depending on the storage format for the integer vector and the degree of sparsity, this could be an efficient scheme. They include a high level description of a machine that uses a floating point pipeline for the arithmetic processor; however, details such as the integer format are not discussed.

We next consider algorithm development for general sparse matrices. In one approach, changes are made in the implementation of standard methods in order to improve performance; in the other approach, different ordering schemes are employed specifically to introduce parallelism. Most of the implementation changes have focused on vector computers, and we begin the discussion with these techniques.

As noted in Duff [1984] (see also Duff [1982a, b]), for example, the difficulty with vectorizing a general sparse routine is the indirect addressing as given below.

```
DO 10    II = I1, I2
         I = INDEX (II)
         T(I) = T(I) + CONST*A(II)
10 CONTINUE
```

This loop may be treated directly using a GATHER operation to form a vector out of the T's, performing the arithmetic operation on the vector, and then using a SCATTER operation to distribute the new vector to the proper locations in T. The Cyber 200 provides hardware instructions for these operations while on the Cray they are available as assembly language routines. For the Cray-1 this technique has an asymptotic rate of 7 MFLOPS or approximately double that obtained from the FORTRAN code given above (see Duff [1984]). Additional results involving assembly coding also are reported in Dodson [1981] and Duff and Reid [1982].

In order to avoid the problem of indirect addressing in sparse systems, Duff [1984] proposed using a frontal technique based on the variable band or profile scheme suggested by Jennings [1966]. The idea is not to form the entire matrix but to eliminate each variable whenever its row and column are available. This allows one to work with a relatively small dense submatrix whose size is governed by the distance from the main diagonal of the first nonzero in a row. The size may vary as the process moves down the diagonal since all elements will not in general be the same distance from the diagonal. No extra storage is used because the factorization produces fill inside the first nonzero of each row. By holding appropriate values in the vector registers in the spirit of the algorithms discussed earlier, Duff [1984] claims performance in the 80 megaflop range for the factorization, a dramatic improvement over general sparse techniques. The frontal technique is particularly attractive for finite element analysis since the factorization may be coupled with the assembly of the global stiffness matrix so that the entire matrix is never formed. The technique also offers a possible solution to the I/O problem produced by very large problems and known to be potentially devastating on high performance systems (see for example Knight, et al. [1975]).

In a series of papers, Calahan [1979a, b], [1981a] has suggested a block approach to solving sparse systems that has some of the characteristics of the frontal technique discussed above. Again the motivation is to reduce the cost of indirect addressing usually associated with sparse methods. In Calahan [1979a], for example, it is pointed out that sparse matrices arising from discretization of partial differential equations typically give rise to matrices that are globally sparse but locally dense. This observation is particularly true if the fill associated with direct methods is taken into account. Motivated by the ability of the Cray to process relatively short vectors efficiently, Calahan [1979a, b] suggests the use of block factorization methods where efficient dense solvers are used to factor the diagonal blocks. As one would expect, the approach becomes very efficient on the Cray as the block size approaches 64. Based on a very accurate simulator described in Orbits [1978], Orbits and Calahan [1978] and Calahan [1979a] predict performance exceeding 100 MFLOPS.

The choice of the blocks is an interesting issue, particularly if the sparse matrix is not sufficiently regular. Calahan [1979a] suggests that the blocking should be done on the LU map of the factored matrix, thus taking into account any fill that may take place. He also proposes that it be based on selecting the largest diagonal block available followed by the next largest and so on. There remains the problem of determining when to end one block and begin with a new one since there is a trade-off between the inclusion of a row in order to approach the optimum size of 64 and the unnecessary computations that may result because of the structure of the row. This is illustrated in Fig. 3.6 where one must decide between the blocking indicated by solid lines and the one indicated by dashed lines. Note that neither this approach nor the frontal method would be as attractive on the Cyber 200 because of the short vector lengths.

$$\begin{bmatrix} x & x & x & & & \\ x & x & 0 & & & \\ x & 0 & x & & & \\ & & & x & x & 0 \\ & & & x & x & 0 \\ & & & 0 & 0 & x \end{bmatrix}$$

FIG. 3.6. *Matrix blocking.*

The matrix in Fig. 3.6 also demonstrates that blocking can be used to introduce parallelism. If the solid line blocking is used, then the 4×4 block cannot be factored until the 2×2 block is factored and the $3, 1$ element is eliminated. However the dashed line blocking decouples the matrix and makes it possible to factor both 3×3 blocks simultaneously. For systems with a sufficiently large number of decoupled diagonal blocks of the same size and structure, this strategy could be effective on the Cyber 200 where vectors would consist of appropriate elements from successive blocks. Arrays of processors could also be used on the system, and if the array is of MIMD-type, the blocks could have a different size and structure.

This suggests another way to seek parallelism in the sparse matrix problem, namely, can the underlying grid be numbered or can the rows and columns be interchanged so as to decouple blocks of the matrix? In an early paper, Calahan [1973] noted that the odd-even reduction strategy of Buneman [1969] applied to tridiagonal matrices could be viewed as a decoupling of those matrices. This will be treated in more detail shortly. Calahan [1973] also discussed a reordering strategy for finding diagonal submatrices in order to introduce parallelism.

Substructuring. The existence of diagonal blocks that are also diagonal matrices, although attractive, is not necessary for parallel factorization of a matrix. Thus we seek orderings which produce diagonal blocks, with no particular structure, that are decoupled from one another. Such an ordering has been used by structural engineers and is called substructuring (see, for example, Noor, et al. [1978], and also Golub and Mayers [1983] and Widlund [1984] for related approaches). The motivation for substructuring in structural analysis was not to introduce parallelism but to decouple as much as possible different parts of a structure that were united by a relatively small number of points; for example, the wing and fuselage of an aircraft would be treated as separate structures joined by a few points where the wings are attached. Conceptually, the situation is depicted in Fig. 3.7, in which the circle points represent interface nodes between the two regions. Notice that the regions may consist of different types of elements—in this case rectangular and triangular elements. The nodes in the region may be numbered in any appropriate order but the interface points are numbered last.

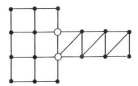

FIG. 3.7 *Substructuring.*

This gives rise to a block matrix of the form

$$\begin{bmatrix} A_1 & & C_1 \\ & A_2 & C_2 \\ D_1 & D_2 & B \end{bmatrix}$$

where the A matrices represent the two substructures, the B matrix represents the interface points, and the C and D matrices represent the dependencies between the interface nodes and the two regions. For many problems, the matrix is symmetric so that $D_i = C_i^T$, $i = 1, 2$. The A matrices may be factored in parallel, and then steps of the form $B - D_1 A_1^{-1} C_1$ are used to eliminate the off-diagonal blocks. Finally the modified B matrix is factored. This process may be generalized to any number of substructures, and is discussed in more detail in Adams and Voigt [1984a]. They use a three-dimensional cube as a model region and obtain formulas to help in the selection of the number of substructures, for if too many are chosen, there will be too many interface nodes and the work involved in factoring the modified B matrix will dominate all other computation. They also compared the technique with a parallel band solver and found that for sufficiently large problems the substructuring technique has advantages.

It should be noted that the nested dissection process described earlier may be viewed as a type of substructuring in which the ordering is chosen so as to minimize storage requirements and operation counts. If the dissection is carried to completion, the diagonal blocks of the resulting matrix reduce to single elements and the upper left-hand corner of the matrix is diagonal. If the dissection is stopped early, as discussed in George, et al. [1978a], the matrix has a block structure of the type obtained by substructuring.

The computing system considered for the substructuring study in Adams and Voigt [1984a] was of MIMD type. This is particularly attractive because in general the A matrices will not be of the same size nor will they have the same structure. Consequently, it would be difficult to use a vector processor where the vectors were defined across substructures or submatrices. For sufficiently large problems, the Cray would be effective applied to each diagonal block in turn; however, the relatively short vector lengths would probably make the technique less desirable for the Cyber 200.

Tridiagonal systems. As we have already pointed out, the degree of parallelism for factorization methods is governed by the semi-bandwidth, m, of the linear system. The tacit assumption has been that m is sufficiently large so that vector operations are efficient or so that there is reasonable processor utilization in a parallel system. However, for small bandwidth systems, and in particular tridiagonal systems ($m = 1$), the methods discussed above are inappropriate, and we will now focus on algorithms which have been designed specifically for tridiagonal systems.

If we consider an LU factorization of a tridiagonal matrix A where L is unit lower bidiagonal and U is upper bidiagonal, the usual algorithm is inherently serial. Defining the ith row of these matrices as $(0, \cdots, 0, c_i, a_i, b_i, 0, \cdots, 0)$, $(0, \cdots, 0, l_i, 1, 0, \cdots, 0)$, and $(0, \cdots, 0, u_i, b_i, 0, \cdots, 0)$ respectively, the ith element of the diagonal of U is given by

$$(3.7) \qquad\qquad u_i = a_i - c_i b_{i-1} / u_{i-1}.$$

Since u_i depends on u_{i-1}, expression (3.7) cannot be evaluated directly using vector operations or an array of processors. This example points out the importance of recurrence relations and indicates why they are a particular problem for parallel processing. We will not discuss algorithms for recurrence relations but instead refer to

Kogge and Stone [1973], Hyafil and Kung [1977], Heller [1978], Kuck [1978] and Hockney and Jesshope [1981] for detailed discussions and additional references.

There are algorithms which avoid the difficulty suggested by (3.7). The first of these was introduced by Stone [1973] and it still represents one of the very few new algorithms that have resulted from considering parallel computation; most others are attempts to introduce parallelism into traditional sequential algorithms. Stone began with the well-known fact that the formulas required by LU factorization of a tridiagonal matrix could be expressed as first and second order linear recurrences. In particular, if one uses the recurrence

$$(3.8) \qquad q_0 = 1, \quad q_1 = a_1, \quad q_i = a_i q_{i-1} - c_i b_{i-1} q_{i-2}, \qquad i = 2, \cdots, n$$

then u_i of (3.7) is given by

$$u_i = q_i / q_{i-1}, \qquad i = 1, \cdots, n.$$

On the surface this does not appear to help, but Stone also observed that the recurrence (3.8) can be written in matrix form as

$$(3.9) \quad Q_i \equiv \begin{bmatrix} q_i \\ q_{i-1} \end{bmatrix} = \begin{bmatrix} a_i & -c_i b_{i-1} \\ 1 & 0 \end{bmatrix} \begin{bmatrix} q_{i-1} \\ q_{i-2} \end{bmatrix} \equiv G_i Q_{i-1} = \left(\prod_{j=2}^{i} G_j \right) Q_1, \qquad i = 2, 3, \cdots.$$

Similar expressions are given for the forward and back substitution.

Stone proposed the parallel computation of (3.9) by recursive doubling, a procedure which, in the simplest case, expresses the $2i$th element in a sequence in terms of the ith. Thus for $n = 2^k$, the nth component can be computed in $\log n$ steps. For purposes of illustration, let $n = 8$ and define $p_{ij} \equiv \prod_{l=i}^{j} d_l$, where d_l, $l = 1, \cdots, n$ is a set of scalars. Then Fig. 3.8 shows the k vector multiply operations that compute each of the p_{ij} for $j = 1, 2, \cdots, n$. The blanks left in some of the vectors are to indicate that no operations are performed there. Thus the first is a multiply of length 7, then 6, and then 4. In general for $n = 2^k$, there are k multiplies of length $n - 2^i$ for $i = 0, 1, 2, \cdots, k-1$. The average length of each multiply is given by

$$n_a = (1/k) \sum_{i=0}^{k-1} (n - 2^i) = (n(\log n - 1) + 1) / \log n \approx n.$$

Since there are $\log n$ such multiplies, the total number of results generated is then $n \log n - n + 1$. Thus we have replaced a serial computation requiring $O(n)$ computations with one that requires $O(n \log n)$ computations. If there are n processors available,

$$\begin{bmatrix} d_1 \\ d_2 \\ d_3 \\ d_4 \\ d_5 \\ d_6 \\ d_7 \\ d_8 \end{bmatrix} * \begin{bmatrix} \\ d_1 \\ d_2 \\ d_3 \\ d_4 \\ d_5 \\ d_6 \\ d_7 \end{bmatrix} = \begin{bmatrix} p_{11} \\ p_{12} \\ p_{23} \\ p_{34} \\ p_{45} \\ p_{56} \\ p_{67} \\ p_{78} \end{bmatrix}, \begin{bmatrix} p_{11} \\ p_{12} \\ p_{23} \\ p_{34} \\ p_{45} \\ p_{56} \\ p_{67} \\ p_{78} \end{bmatrix} * \begin{bmatrix} \\ \\ p_{11} \\ p_{12} \\ p_{23} \\ p_{34} \\ p_{45} \\ p_{56} \end{bmatrix} = \begin{bmatrix} p_{11} \\ p_{12} \\ p_{13} \\ p_{14} \\ p_{25} \\ p_{36} \\ p_{47} \\ p_{58} \end{bmatrix}, \begin{bmatrix} p_{11} \\ p_{12} \\ p_{13} \\ p_{14} \\ p_{25} \\ p_{36} \\ p_{47} \\ p_{58} \end{bmatrix} * \begin{bmatrix} \\ \\ \\ \\ p_{11} \\ p_{12} \\ p_{13} \\ p_{14} \end{bmatrix} = \begin{bmatrix} p_{11} \\ p_{12} \\ p_{13} \\ p_{14} \\ p_{15} \\ p_{16} \\ p_{17} \\ p_{18} \end{bmatrix}$$

FIG. 3.8. *Recursive doubling.*

then we have gone from $O(n)$ steps to $O(\log n)$, a clear improvement. However, as was pointed out in Lambiotte and Voigt [1975], for vector computers the total number of operations is important, so that even though vector operations can be used, at some point the $n \log n$ operations will dominate and the vector algorithm will require more time than the scalar algorithm. This led them to propose the following definition for a consistent algorithm. A vector implementation of an algorithm for solving a problem of size n is said to be *consistent* if the number of arithmetic operations required by this implementation as a function of n is the same order of magnitude as required by the usual implementation on a serial computer. Thus both the recursive doubling algorithm and the tridiagonal algorithm which uses it are inconsistent. Stone [1975] and Lambiotte and Voigt [1975] give consistent versions of Stone's original algorithm although the latter paper points out that the asymptotic superiority has little significance for problems whose size might be of practical interest.

Cyclic reduction. Another consistent algorithm known as odd-even reduction or cyclic reduction appears to be the most popular alternative to the standard sequential algorithm. Cyclic reduction was originally proposed by Gene Golub and Roger Hockney and is discussed in Hockney [1965] for the block tridiagonal systems arising from the 5-point difference approximation for Poisson's equation. Subsequently, several authors, including Hockney [1970] and Ericksen [1972], pointed out that the algorithm could also be adapted to general tridiagonal systems. The idea is to eliminate the odd numbered variables in the even numbered equations by performing elementary row operations. Thus if $R(2i)$ represents the $2i$th row of the tridiagonal matrix, the following operations can be performed in parallel for $i = 1, \cdots, (n-1)/2$, assuming n is odd:

$$(3.10) \qquad R(2i) - (c_{2i}/a_{2i-1}) * R(2i-1) - (b_{2i}/a_{2i+1}) * R(2i+1).$$

There are several observations about cyclic reduction that should be noted. If the matrix is stored by diagonals, then expression (3.10) may be evaluated using vector operations on a computer like the Cray or the Cyber 200. After the step indicated by (3.10) is completed, the resulting system under a reordering is again tridiagonal but only half as large. Thus the process may be continued for k steps until, in the case that $n = 2^k - 1$, only one equation remains; then all of the unknowns are recovered in a back substitution process. The details of these observations are given in Lambiotte and Voigt [1975], where it is also shown that cyclic reduction requires $O(n)$ operations and is thus consistent. It should be noted that this is another example of the paradigm of reordering to increase parallelism that was discussed in §1.

One major difficulty with cyclic reduction is that it can require significant data rearrangement between steps. For example, on the Cyber 200 one cannot apply vector operations directly to every other element of a vector. Thus extra operations must be employed to reformat those elements into a new vector. Lambiotte and Voigt [1975] show that this overhead accounts for approximately half of the total operations. Their analysis is based on STAR-100 timing but the situation remains essentially the same for the Cyber 200. Accessing elements of a vector on the Cray at a fixed increment or stride is possible but it may lead to a degradation in performance if the same memory bank is read too frequently. This was recognized in Kershaw [1982], where a storage scheme is discussed that makes it possible to avoid memory bank conflicts. Results reported there indicate that the algorithm is more than six times faster than the scalar algorithm for matrices of order $n > 1000$; even for small systems with $n \sim 10$ the cyclic reduction algorithm is faster. The importance of this overhead was also discussed by Boris [1976b] who considered an implementation of cyclic reduction for the TI ASC, a computer which did not require that a vector be defined as elements in contiguous memory locations.

Because of the overhead of data rearrangement, one would expect that for sufficiently small n the serial algorithm would run faster than cyclic reduction. This leads to the possibility of a polyalgorithm in which cyclic reduction is used until the matrix size is reduced to the point that the serial algorithm is more efficient. This idea is discussed in Madsen and Rodrigue [1976], where it is shown to be superior to an inconsistent algorithm proposed by Jordan [1974]. The idea also serves as a basis for discussion of many algorithms in Hockney and Jesshope [1981], including those for the tridiagonal problem.

Under appropriate assumptions, Heller [1976] showed that during the cyclic reduction process the off-diagonal elements decrease in size relative to the diagonal entries at a quadratic rate. This means that it may be possible to teminate the process in less than $\log n$ steps. For vector computers it is thus possible to avoid the last few steps which are with short vectors; for parallel computers it means that poor processor utilization associated with the last few steps may be avoided. A similar phenomenon was observed earlier by Malcolm and Palmer [1974] for an LU factorization algorithm for tridiagonal systems which are real, symmetric and diagonally dominant with constant diagonals. Their idea was used by O'Donnell, et al. [1983] as a basis for a fast Poisson solver tailored for the Floating Point Systems, Inc. FPS-164.

Cyclic reduction for block tridiagonal matrices has been studied for parallel computers by Kapur and Browne [1981], [1984] who consider implementations on the TRAC computer. They also considered a variant of cyclic reduction known as odd-even or cyclic elimination that was introduced in Heller [1976], [1978]. In this elimination method expression (3.10) is applied to *each* equation (or block) rather than to just the even ones. The result is that the off-diagonal entries move away from the diagonal so that after $\log n$ steps a diagonal matrix remains and the solution is obtained immediately without a back substitution process. As with cyclic reduction, the off-diagonal elements decrease at a quadratic rate making early termination an attractive alternative. The algorithm is inconsistent, requiring $O(n \log n)$ operations. However, it was superior to cyclic reduction on the TRAC. This is made possible because the extra operations of the elimination method are done in parallel at no extra cost and because there is no back substitution step. Thus we have an example of a good uniprocessor algorithm being outperformed by a poor uniprocessor algorithm in a parallel environment. Another interesting aspect of their study is that they were able to implement the algorithm so that the overhead cost of data movement, synchronization, etc. was kept to approximately ten percent of the total time. Gannon, et al. [1983] also recognized the potential superiority of odd-even elimination in their study of implementing parallel algorithms on the CHiP systems. The algorithm was also used by Gannon and Panetta [1985] in a study of the performance of the SIMPLE code on CHiP. In a recent paper, Johnsson [1984b] gives a thorough analysis of the implementation of cyclic reduction and some variants on a family of parallel computers called ensemble architectures. These designs are of MIMD type using simple processors and no global memory. A variety of interconnection schemes are considered.

Other tridiagonal methods and stability. To this point we have said nothing about the stability of the tridiagonal schemes. There has been the tacit assumption, for example, that no pivoting is required, and in fact there does not appear to be any way to incorporate a pivoting strategy into the algorithms discussed. Several authors have noted that cyclic reduction is just Gaussian elimination applied to PAP^T for a particular permutation matrix P (see, for example, Lambiotte and Voigt [1975]). Thus the algorithm is numerically stable for matrices for which Gaussian elimination is stable without pivoting, for example, symmetric positive definite or diagonally dominant matrices. The situation is not as attractive for Stone's algorithm. Using a stability analysis technique for recurrence relations introduced in Sameh and Kuck [1977a], Dubois and Rodrigue [1977a] have shown that the algorithm is in general unstable, suffering from exponential error growth.

As discussed earlier in this section, Givens transformations may be used to overcome the difficulties of pivoting for stability. Sameh and Kuck [1978] present two such algorithms for tridiagonal systems using $O(n)$ processors. One of the algorithms requires $\log n$ steps but can suffer from exponential growth of errors; the more stable version requires $O[(\log \log n)(\log n)]$ steps. Another Givens based algorithm is discussed in Hatzopoulos [1982]. The different feature of this algorithm is that the Givens transformations are applied from the top and from the bottom of the matrix simultaneously, thus increasing the degree of parallelism by a factor of two but still requiring $O(\log n)$ steps on $O(n)$ processors. Hatzopoulos [1982] also considers using the QIF method discussed earlier in this section. Again the implementation requires $O(\log n)$ steps on $O(n)$ processors. Unfortunately all of these algorithms are inconsistent and unless stability is a problem, they would not be attractive for vector computers. There appears to be no implementation of a Givens transformation based algorithm that is consistent.

There are consistent algorithms other than cyclic reduction. Swarztrauber [1979a, b] introduced an algorithm for tridiagonal systems based on an efficient implementation of Cramer's rule. The algorithm requires $O(\log n)$ steps on $O(n)$ processors but only $O(n)$ total operations are performed. The algorithm also requires only a single divide, and unlike cyclic reduction it is well defined for general nonsingular systems. There has been no formal stability analysis but Swarztrauber reports results comparable to Gaussian elimination with partial pivoting for a series of experiments run on the Cray-1. The algorithm has a slightly higher operation count than cyclic reduction but it is more efficient than Gaussian elimination on the Cray-1 when n exceeds 32. Kascic [1984a] has compared cyclic reduction with the Cramer's rule algorithm and found cyclic reduction to be about twice as fast on the Cyber 205.

A variety of other algorithms have been proposed for tridiagonal systems. For example, Sameh [1981] and Kowalik, Lord and Kumar [1984] consider a block algorithm where the number of blocks is chosen to match the number of processors available. An elimination scheme is used within each block until a reduced system remains. Following an order of elimination suggested by Wang [1981], Kowalik, Lord and Kumar [1984] obtain a system of p equations that must be solved sequentially, where p is the number of blocks. They present results from an implementation on the Denelcor HEP and note that the speedup falls considerably short of p because of extra computation that the algorithm requires. Sameh [1981] considers his algorithm for a linear array of processors. The sequential part of the algorithm is in the back substitution but the algorithm is structured nicely for a linear array so that the communication should not be a major overhead. He shows that for $p = n$ the algorithm requires $O(n^{1/2})$ time including communication.

Because of their inherent parallelism, iterative methods have been considered by Traub [1974b] for solving tridiagonal systems, and further studied by Lambiotte and Voigt [1975] and Heller, et al. [1976]. Traub's idea was to turn the three basic recurrence relations associated with the LU factorization into iterations. For example, equation (3.7) is formulated as

$$u_i^{(k+1)} = a_i - c_i b_{i-1}/u_{i-1}^{(k)}, \qquad k = 1, \cdots, m,$$

and the rate of convergence depends on the degree of diagonal dominance of the system. Except for certain situations such as a very strongly diagonally dominant system or where an excellent starting value need only be improved by a few digits, these methods do not appear to be competitive with direct methods such as cyclic reduction.

Most of the work discussed to this point has focused on reducing the parallel computational complexity of the algorithm with occasional concerns for the overhead arising from such things as communication. A rather different approach is taken by Merriam [1985] where, motivated by the small memory that was available on the Illiac IV, he considers minimizing the total time to solve tridiagonal systems by trading off extra computation with storage that might require expensive I/O operations. His idea is to save a few carefully selected values from the factorization stage, and then begin the back substitution. When an element is required that is not available it is recomputed using the values saved from the factorization. This idea can have merit in any situation where the cost of communication is high relative to computation.

There are, of course, nontridiagonal matrices of interest whose bandwidth is too small for efficient use of banded solvers on vector or parallel computers. One way to treat these problems is to view them as block tridiagonal and apply block cyclic reduction as discussed in Lambiotte [1975] and Heller [1976]. It would be more attractive to apply the cyclic reduction idea directly, as suggested by Rodrigue, et al. [1976] and Madsen and Rodrigue [1977], so that the parallelism obtained is by the diagonals of the matrix rather than the band. Unfortunately, the numerical stability of the algorithm remains in doubt; indeed, even reasonable conditions on the matrix that guarantee that the algorithm remains well defined (i.e. division by zero cannot occur) have not been given although some numerical experience in Madsen and Rodrigue [1977] did not expose any problems.

Fast Poisson solvers. So far in this section, we have made few assumptions about the differential equations which give rise to the linear systems to be solved. However, for separable problems there are special methods, generally known as "fast" methods, which are considerably better than other direct or iterative methods. These methods are reviewed for scalar computers in, for example, Dorr [1970], Swarztrauber [1977], and Temperton [1979a,b],[1980]. Although some of the algorithms are applicable to more general problems we will again use the Poisson equation on a square in order to provide a focus for the discussion. For this problem, Swarztrauber [1977] has shown how to handle periodic, Dirichlet–Dirichlet, Dirichlet–Neumann, and Neumann–Neumann boundary conditions.

The algorithms to be discussed depend on the Fast Fourier Transform (FFT). Swarztrauber [1982], [1984] contains a thorough discussion of FFT's on vector computers, particularly the Cray, while the Cyber 200 motivated Korn and Lambiotte [1979] and Lambiotte [1979] to develop algorithms for multiple transforms that give rise to vector lengths that are longer than that provided by a single transform. These algorithms maintain the serial complexity of $O(mn \log n)$ for m transforms of length n and exhibit a degree of parallelism of m or n.

Pease [1968], Stone [1971] and Jesshope [1980a] have considered the efficiency of FFT algorithms on a variety of parallel arrays. Since the algorithms depend on data distributed over the entire array rather than on data contained in neighboring processors, communication becomes a significant issue. Finally, Hockney and Jesshope [1981] provide a lengthy discussion of algorithms for both arrays and vector processors including guidelines on the choice of methods depending on the architecture and the size and number of transforms required.

Both Pease and Stone noted that an interconnection scheme known as the perfect shuffle provided the kind of data transmission required by the FFT. For processors P_i, $i=0,\cdots,N-1$, the perfect shuffle provides direct communication as follows:

$$P_i \to P_{2i}, \qquad 0 \leq i \leq \frac{N}{2}-1,$$

$$P_i \to P_{2i+1-N}, \qquad \frac{N}{2} \leq i \leq N-1.$$

This accomplishes an interleaving of transmitted information analogous to what one obtains with a perfect shuffle of a deck of cards. Figure 3.9 shows the perfect shuffle interconnection for eight processors.

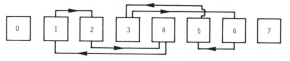

FIG. 3.9. *Perfect shuffle interconnection*.

We now show how the Fast Fourier Transform plays a crucial role in various fast Poisson solvers. Following Hockney and Jesshope [1981], we assume that a 5-point difference scheme is used to discretize the Poisson equation

$$(3.11) \qquad \Delta u = f$$

with doubly periodic boundary conditions on an $N \times N$ grid. Then taking a Fourier transform in the x direction followed by one in the y direction gives rise to the following expression for the transform of the right-hand side f_{jk},

$$\tilde{f}_{jk} = \frac{1}{N} \sum_{q=0}^{N-1} \left(\frac{1}{N} \sum_{p=0}^{N-1} f_{p,q} \exp\left[-\frac{2\pi ijp}{N} \right] \right) \exp\left[-\frac{2\pi ikq}{N} \right], \qquad 1 \leq j, k \leq N.$$

A division of each $\tilde{f}_{jk}$ by the coefficient of the transformed left-hand side of the discrete form of (3.11) gives a value for each transformed unknown variable $\tilde{U}_{j,k}$ and finally the solution $U_{j,k}$ is obtained by an inverse transform of $\tilde{U}_{j,k}$. The serial complexity of the algorithm is $O(N^2 \log N)$ and the degree of parallelism in calculating $\tilde{f}_{j,k}$ is either N, if a serial algorithm is used on N transforms in parallel, or N^2 if a parallel algorithm is used. As usual, the choice would depend on the machine and the value of N.

Another algorithm based on the FFT, which also has serial complexity of $O(N^2 \log N)$, is known as matrix decomposition. It was introduced by Buzbee, et al. [1970] and was first analyzed as a parallel algorithm by Buzbee [1973]. If (3.11) is discretized using the 5-point difference formula on an evenly spaced square grid one obtains the block tridiagonal system

$$(3.12) \qquad \begin{bmatrix} A & I & & & \\ I & A & \cdot & \cdot & \\ & \cdot & \cdot & \cdot & \\ & \cdot & \cdot & \cdot & I \\ & & I & \cdot & A \end{bmatrix} \begin{bmatrix} U_1 \\ \vdots \\ \vdots \\ U_N \end{bmatrix} = \begin{bmatrix} f_1 \\ \vdots \\ \vdots \\ f_N \end{bmatrix}$$

where A is an $N \times N$ tridiagonal matrix whose ith row is $(0, \cdots, 0, 1, -4, 1, 0, \cdots, 0)$, and U_i and f_i are N-vectors. Matrix decomposition is based on the fact that the eigensystem of A is known explicitly and thus the factorization $V^T A V = \Lambda$ is possible where Λ is a diagonal matrix whose entries are the eigenvalues of A. Using this relationship, (3.12) may be rewritten as

$$\Lambda \tilde{U}_1 + \tilde{U}_2 = \tilde{f}_1,$$

(3.13) $\qquad\qquad \tilde{U}_{i-1} + \Lambda \tilde{U}_i + \tilde{U}_{i+1} = \tilde{f}_i, \qquad i = 2, \cdots, N-1,$

$$\tilde{U}_{N-1} + \Lambda \tilde{U}_N = \tilde{f}_N,$$

where $\tilde{U}_i = VU_i$ and $\tilde{f}_i = Vf_i, i = 1, \cdots, N$. Because of the form of the eigenvectors of A, $\tilde{f}_i$ may be computed using a fast sine transform. Then $\tilde{U}_i$ is obtained by solving the systems that result from reorganizing (3.13) into N independent tridiagonal systems each one of which depends on a single eigenvalue of A. The solution U_i is recovered from $\tilde{U}_i$ by means of an inverse sine transform. Thus the degree of parallelism and the appropriateness of the method for a particular computer depend on algorithms for the FFT and for systems of tridiagonal equations.

Sameh, et al. [1976] obtained a complexity of $O(\log N)$ parallel steps for matrix decomposition on a parallel computer consisting of N^2 processors with an arbitrarily powerful interconnection network that required unit time for the transfer of a piece of data from any processor to any other processor. The interconnection requirement could be relaxed to a perfect shuffle network without serious degradation in performance. Sameh [1984a] has also considered a ring of p processors, $p < N$, where each processor can simultaneously perform an arithmetic operation, receive a floating point number from an immediate neighbor and transmit a floating point number to another neighbor. With such a system, he shows that matrix decomposition requires $O((N^2/p) \log N)$ parallel steps, resulting in a speedup of $O(p)$. He also considers a three-dimensional problem using a 7-point difference approximation on an N^3 grid. By using N copies of the ring of N processors with each ring attached to a global memory, it is possible to solve the Poisson equation in $O(N \log N)$ parallel steps. Performance degrades linearly for $r < N$ rings of $p < N$ processors.

Vajtersic [1982] reports results obtained from an implementation of matrix decomposition on the MIMD system EGPA (Erlangen General Purpose Array) under development at the Erlangen–Nurnberg University. The system consisted of a pyramid of five processors with four processors serving as slaves to the apex processor. Using the four processors for the execution of the algorithm he obtained speedups increasing to 3.6 for N ranging to 128. The speedup figures are not as high as they might be because the full parallelism of the algorithm is not utilized in order to simplify synchronization and data transfer.

As mentioned earlier in the discussion of tridiagonal systems, cyclic reduction was developed as a fast Poisson solver for a system of the form (3.12) on which it exhibits serial complexity of $O(N^2 \log N)$. Its parallel implementation is analogous to that discussed for the tridiagonal problem; details for the Illiac IV are given in Ericksen [1972].

Hockney [1965] used one step of cyclic reduction and then solved the remaining system, which is half the original size, by matrix decomposition, resulting in an algorithm he called FACR. Later, Hockney [1970] noted that the overall work could be reduced by taking more cyclic reduction steps. The algorithm known as FACR(l) involves l steps of cyclic reduction, the resulting system is solved by matrix decomposition, and the solution of the discretized Poisson equation is obtained by a back substitution step. Swarztrauber [1977] showed that the minimum computational complexity of $O(N^2 \log \log N)$ is obtained with $l = \log \log N$.

Grosch [1979b] presents an analysis of the FACR(l) algorithm including communication costs for arrays with a nearest neighbor connection and a nearest neighbor connection augmented with a perfect shuffle. He found that the augmented array would operate with an efficiency of around 90 percent for a wide range of values for N while the efficiency of the other array would drop off rapidly with increasing N.

As was noted in the previous discussion of cyclic reduction for tridiagonal systems, the degree of parallelism decreases as more steps are taken and Temperton [1980] points this out for the FACR(l) algorithm. This phenomenon has prompted research on the selection of the appropriate value of l to maximize performance on a variety of vector and parallel computers (see, for example, Hockney and Jesshope [1981] and Hockney [1982a], [1983b] and the references therein). In particular, evaluating the FFT more rapidly will lead to smaller values of l while solving tridiagonal systems faster will lead to a larger value of l. The algorithm has been used on a variety of computers including the Illiac IV (Ericksen [1972]), the Cray-1 (Temperton [1979b]), and the Cyber 205 and ICL DAP (Hockney [1983b]).

Finally, we should mention that the multigrid algorithm can also be viewed as a fast Poisson solver because of its theoretical complexity of $O(N^2)$. We defer a detailed discussion until the next section since the method is appropriate for more general partial differential equations but point out that Grosch [1979b] has studied parallel implementation issues for a Poisson solver that indicates that the method is attractive on arrays of processors.

4. Iterative and time marching methods. The parallel implementation of most of the usual iterative methods for discrete elliptic equations has been studied extensively by a number of authors. Some of the earlier papers were Ericksen [1972], Hayes [1974], and Lambiotte [1975], primarily for the Illiac IV, the TI-ASC, and the CDC STAR-100, respectively, and Morice [1972] for general parallel processors. We also note the papers by Heller [1978] and Ortega and Voigt [1977], which survey many aspects of iterative methods for vector and parallel computers up to that time. More recent surveys which include material on iterative methods are Buzbee [1981], [1983a], Evans [1982b], Feilmeier [1982], Hockney and Jesshope [1981] and Sameh [1983].

For simplicity and ease of presentation much of our discussion will be for the model problem of Laplace's equation on a square with Dirichlet boundary data, discretized using the 5-point difference star. Such a problem would, of course, actually be solved by one of the fast Poisson solvers mentioned in the last section but it makes a convenient example with which to treat many of the issues that arise in more general problems.

Jacobi's method. The discrete domain is shown in Fig. 4.1 where the boundary points are indicated by b's, and N is the number of interior points in each row and column. The classical Jacobi method

$$(4.1) \qquad u_{i,j}^{k+1} = \frac{1}{4} \left[u_{i+1,j}^k + u_{i-1,j}^k + u_{i,j+1}^k + u_{i,j-1}^k \right],$$

where the superscript denotes the iteration number, and u_{ij} is the solution at the i,j grid point, is generally considered to be a prototype parallel method. However, care is needed in certain aspects of its implementation in order to achieve the greatest degree of parallelization. For example, (4.1) would be more efficiently implemented on the

$$
\begin{array}{llllll}
b & b & b\ b\ b\ b & & & \\
 & & & (N+2)^2 & & \\
b & \cdot & \cdot\ \cdot\ \cdot & b & & \\
b & \cdot & \cdot\ \cdot\ \cdot & b & & \\
b & \cdot & \cdot\ \cdot\ \cdot & b & & \\
2N+5 & & & & & \\
b & \cdot & \cdot\ \cdot\ \cdot & b & & \\
N+3 & N+4 & & & & \\
b & b & b\ b\ b\ b & & & \\
1 & 2 & & N+2 & &
\end{array}
$$

Fɪɢ. 4.1. *Grid points.*

Cray in a row by row fashion if N were 64 or a multiple thereof and on a $p \times p$ array of processors if N were a multiple of p.

On the Cyber 200 machines, on the other hand, we would like the vector lengths to be as long as possible. One could carry out (4.1) row by row, but then the vectors would only be of length N and one would pay N times the number of start-up penalties. Alternatively, we can use vectors of length $O(N^2)$ by treating the boundary positions as unknowns. That is, let U now denote an $(N+2)^2$ long one-dimensional array with the lexicographic ordering of Figure 4.1, and use the notation $U(K; L)$ to denote the L-long subvector starting at the Kth position of U. With $M1 = (N+1)(N+2)-1$ and $M2 = N(N+2)-2$, we can then implement (4.1) by the instructions

(4.2)
$$
T(2; M1) = U(2; M1) \overset{+}{_2} U(N+3; M1),
$$
$$
U(N+4; M2) = T(2; M2) \overset{+}{_2} T(N+5; M2)
$$

where T is a temporary vector and where we have used $\overset{+}{_2}$ to denote an "average" instruction, that is, addition followed by division by 2. Such an instruction is available on the Cyber 200's and takes essentially the same time as an addition.

As a penalty for using vectors whose length is the total number of grid points, the final instruction of (4.2) will overwrite the positions $2N+4$, $2N+5$, $3N+6$, $3N+7, \cdots$ corresponding to boundary positions along the vertical sides, thus destroying the correct boundary values. One would then have to restore these values before the next iteration. On the Cyber 200's however, there is a convenient feature which permits storage to be controlled by a bit vector (the control vector); this can be used to ensure that the boundary positions are not overwritten and, hence, no "fixing up" is needed before the next iteration. The instruction time is no greater using the control vector but, of course, one pays the penalty for storage of approximately $N^2/64$ words for the bit vector.

Although the Jacobi method vectorizes well, it has not been used in practice because of its slow convergence. However, Schonauer [1983a, b] has reported promising results on a "meander" Jacobi overrelaxed method in which the relaxation parameter varies with the iteration number in rather complicated ways.

SOR. Whereas the Jacobi iteration is often cited as a "perfect" parallel algorithm, the Gauss–Seidel and SOR iterations are considered to be the opposite. The usual serial code for Gauss–Seidel in the context of (4.1) would have new values at each point replace the old as soon as they are computed; it is this recursive process that is not amenable to vectorization. However, several early authors (e.g. Ericksen [1972], Hayes [1974], Lambiotte [1975]) observed that by using the classical red-black ordering of the grid points, as shown in Fig. 4.2, Gauss–Seidel can be carried out in the same fashion as the Jacobi iteration by using two vectors of length $O(N^2/2)$ corresponding to the red and black points. The boundary points would be handled in the same way as with Jacobi's method and the introduction of the SOR parameter causes no difficulty. The time per iteration for SOR carried out in this way should be little more than for the Jacobi iteration so that the SOR method is potentially very useful for parallel computation. Lambiotte [1975] also considered a diagonal ordering for the grid points but showed that this is inferior to the red-black ordering.

$$
\begin{array}{ccccc}
\cdot & \cdot & \cdot & \cdot & \cdot \\
 & B11 & R9 & B12 & \\
\cdot & \cdot & \cdot & \cdot & \cdot \\
B8 & R7 & B9 & R8 & B10 \\
\cdot & \cdot & \cdot & \cdot & \cdot \\
R4 & B6 & R5 & B7 & R6 \\
\cdot & \cdot & \cdot & \cdot & \cdot \\
B3 & R2 & B4 & R3 & B5 \\
\cdot & \cdot & \cdot & \cdot & \cdot \\
 & B1 & R1 & B2 & \\
\end{array}
$$

FIG 4.2. *The red-black ordering.*

In related work, Buzbee, et al. [1977] discussed the treatment of the equation $(\alpha u_x)_x + (\beta u_y)_y = f$ on the Cray-1. They used the 5-point difference star with the red-black ordering, and also considered a skewed 5-point difference scheme using the NE, SE, SW and NW grid points rather than the usual north, south, east, west ones. In the latter scheme, they ordered the grid points red-black by columns.

While the red-black ordering allows an efficient implementation of the SOR method for the 5-point difference scheme, it does not work for higher order finite difference or finite element discretizations or for more general elliptic equations which contain mixed partial derivative terms. However, several authors have observed that the red-black ordering can be extended to a "multicolor" ordering which can give the same effect as the red-black ordering for the 5-point star. Hotovy and Dickson [1979] used a three-color ordering for a finite difference approximation of the small disturbance equation of transonic flow, Hackbusch [1978] used a four-color ordering for 9-point finite difference stars and Adams and Ortega [1982] gave a general treatment of the idea which we now describe (we note that Young [1971] had much earlier used a three-color ordering but not in the context of parallel computing and that Berger, et al. [1982] use multicolor orderings for the assembly of finite element equations).

The basic idea of the multicolor ordering is to label (color) the grid points in such a way that there is local decoupling between the unknowns. This leads to the system being expressed in the form

$$
(4.3) \quad
\begin{bmatrix}
D_1 & B_{12} & \cdot & \cdot & \cdot & B_{1p} \\
B_{21} & D_2 & & & & \\
\cdot & & \cdot & \cdot & & \vdots \\
\cdot & & \cdot & \cdot & \cdot & \\
\cdot & & & \cdot & \cdot & B_{p-1,p} \\
B_{p1} & \cdot & \cdot & \cdot & B_{p,p-1} & D_p
\end{bmatrix}
\begin{bmatrix}
x_1 \\
\vdots \\
\vdots \\
x_p
\end{bmatrix}
=
\begin{bmatrix}
b_1 \\
\vdots \\
\vdots \\
b_p
\end{bmatrix}
$$

for p colors, where the D_i are diagonal. The case $p = 2$ is the red-black ordering. The $(k+1)$st Gauss–Seidel iterate is then

$$D_i \mathbf{x}_i^{k+1} = -\sum_{j<i} B_{ij}\mathbf{x}_j^{k+1} - \sum_{j>i} B_{ij}\mathbf{x}_i^k + \mathbf{b}_i, \qquad i = 1, \cdots, p$$

and is effectively implemented by p Jacobi sweeps. As an example, Fig. 4.4 gives a coloring suitable for a discretization in which each grid point is coupled to its eight nearest neighbors as indicated in Fig. 4.3.

FIG. 4.3. *Eight nearest neighbors.*

A variety of other examples could be given (see, e.g., Adams [1982]). Provided that the domain of the differential equation is a rectangle or some other regular two- or three-dimensional region and that the discretization pattern is repeated at each grid point, it is usually evident how to color the points to achieve the desired result. However, obtaining the minimum number of colors for arbitrary discretizations is equivalent to the graph coloring problem which is *NP*-complete.

$$
\begin{array}{cccccccc}
\cdot & \cdot & \cdot & \cdot & \cdot & \cdot & \cdot & \cdot \\
R & B & W & O & R & B & W & 0 \\
\cdot & \cdot & \cdot & \cdot & \cdot & \cdot & \cdot & \cdot \\
W & O & R & B & W & O & R & B \\
\cdot & \cdot & \cdot & \cdot & \cdot & \cdot & \cdot & \cdot \\
R & B & W & O & R & B & W & O \\
\cdot & \cdot & \cdot & \cdot & \cdot & \cdot & \cdot & \cdot \\
W & O & R & B & W & O & R & B \\
\end{array}
$$

FIG. 4.4. *Four-color ordering of the grid points.*

Other orderings which are conducive to parallel computing have also been used. O'Leary [1984] gives a number of interesting orderings, one of which is illustrated in Fig. 4.5. Here, the nodes are grouped in blocks of five, except at the boundaries. First, all points labeled 1 are ordered, followed by all points labeled 2, then all points labeled 3. The resulting system has the form (4.3) with $p = 3$ but now the D_i are block diagonal matrices with blocks that are 5×5 or less. A block SOR iteration can then be carried out with block Jacobi sweeps which involves solving 5×5, or smaller, systems.

$$
\begin{array}{cccccccccc}
3 & 3 & 1 & 1 & 3 & 3 & 1 & 2 & 3 & 3 \\
3 & 3 & 2 & 2 & 3 & 3 & 2 & 2 & 3 & 1 \\
3 & 1 & 2 & 2 & 1 & 1 & 2 & 2 & 1 & 1 \\
1 & 1 & 2 & 3 & 1 & 1 & 3 & 3 & 1 & 1 \\
1 & 1 & 3 & 3 & 1 & 2 & 3 & 3 & 2 & 2 \\
\end{array}
$$

FIG 4.5. *O'Leary's P_3 ordering.*

There are at present only partial results concerning the rate of convergence of SOR using multicolor and related orderings. Adams [1982] showed that, in general, these are not consistent orderings in the sense of Young. However, O'Leary [1984] proved that if the matrix (4.3) is an irreducible Stieltjes matrix then the asymptotic rate of convergence is no worse than for the natural ordering. A different approach was motivated by the implementation of SOR on the Denelcor HEP by Patel and Jordan [1985] in which they use the natural ordering and let the updating take place as soon as the requisite values at the neighboring points are available. This is illustrated in Fig. 4.6 for the 9-point stencil of Fig. 4.3; each sequence of numbers represents a grid point and indicates the times at which the corresponding unknown can be updated. Further analysis of this procedure by Adams and Jordan [1984] showed that it is equivalent, in certain cases and up to transient effects, to carrying out SOR under a multicolor ordering. In particular, they show for a wide class of discretization stencils that the spectral radius of the SOR iteration matrix for certain multicolor orderings is the same as that of the iteration matrix for the natural rowwise ordering and hence, in these cases, the asymptotic rate of convergence for the multicolor orderings is the same as that for the natural ordering.

11,15,19,23	12,16,20,24	13,17,21,25	14,18,22,26	15,19,23,27
9,13,17,21	10,14,18,22	11,15,19,23	12,16,20,24	13,17,21,25
7,11,15,19	8,12,16,20	9,13,17,21	10,14,18,22	11,15,19,23
5,9,13,17	6,10,14,18	7,11,15,19	8,12,16,20	9,13,17,21
3,7,11,15	4,8,12,16	5,9,13,17	6,10,14,18	7,11,15,19
1,5,9,13	2,6,10,14	3,7,11,15	4,8,12,16	5,9,13,17

FIG. 4.6. *Update times for 9-point stencil.*

We turn now to the implementation of the Jacobi or SOR iterations on an array of processors. As discussed in §2, this will require a suitable distribution of the work amongst the processors so as to minimize processor idleness. In order to carry out these iterations in their mathematical form we also need to ensure that the processors are synchronized before the beginning of each iteration, or in the case of the multicolor SOR method, before the beginning of each Jacobi sweep. This synchronization can be carried out in a number of ways but, in essence, it requires that each processor wait after completion of its part of the computation until all processors have completed their work and the next iteration can begin. This adds two forms of overhead to the computation: one is the work required to verify that every processor is ready for the next iteration; the other is the idle time that some processors may experience while waiting for all processors to complete their tasks.

An alternative that has special appeal in the case of the Jacobi or SOR iterations is to let the processors run asynchronously. This idea goes back at least to the chaotic relaxation methods of Chazan and Miranker [1969] and has been studied in some detail by Baudet [1978], following work of Kung [1976]. See also Deminet [1982], who gives some performance results on the *Cm**, Barlow and Evans [1982], and Dubois and Briggs [1982]. In its simplest form, in the present context, we would simply not worry about synchronization at each iteration in carrying out the Jacobi sweeps of the multicolor SOR method; that is, at any given time, the Jacobi iteration would take as its data the most recently updated values of the variables. In general, this asynchronous iteration will deviate from the synchronized one but this need not diminish the rate of convergence.

A different approach to the parallelization of the SOR iteration is taken by Evans and Sojoodi-Haghighi [1982] who develop methods related to the QIF factorizations of the previous section; they are not SOR methods but have somewhat the same spirit. Let

$A = X - W - Z$ be a splitting of A where the zero/nonzero patterns of X, W and Z are indicated by

$$(4.4) \qquad X = \begin{bmatrix} \boxtimes \end{bmatrix}, \qquad W = \begin{bmatrix} \blacktriangleright \blacktriangleleft \end{bmatrix}, \qquad Z = \begin{bmatrix} \blacktriangledown \\ \blacktriangle \end{bmatrix},$$

that is, X has nonzero elements only on the main diagonal and the cross diagonal, and W and Z have zeros on the main and cross diagonals and nonzeros only in the indicated hatched positions. Evans and Sojoodi-Haghighi then define overrelaxed Jacobi and SOR-like methods by

$$(4.5a) \qquad \mathbf{u}^{k+1} = \left[\omega X^{-1}(W+Z) + (1-\omega)I\right]\mathbf{u}^k + \mathbf{d}, \qquad k = 0, 1, \cdots,$$

$$(4.5b) \qquad \mathbf{u}^{k+1} = (X - \omega W)^{-1}\left[\omega Z + (1-\omega)X\right]\mathbf{u}^k + \mathbf{d}, \qquad k = 0, 1, \cdots$$

called, respectively, the Jacobi-Overrelaxed Quadrant Interlocking (JOQI) and the Successive Overrelaxed Quadrant Interlocking (SOQI) methods. They prove most of the usual types of convergence theorems for these methods; for example, if A is irreducibly diagonally dominant, then SOQI converges for $0 < \omega \leq 1$. If A and X are symmetric positive definite, then SOQI converges for $0 < \omega < 2$ and JOQI converges if and only if $2\omega^{-1}X - A$ is positive definite. The solution of the systems with either X or $X - \omega W$ as the coefficient matrix in (4.5) can be effected by solving $n/2$ 2×2 systems, which can be done in parallel. Since the right-hand sides of these systems can be evaluated in parallel, the degree of parallelism of the methods is essentially the same as that of SOR with the red-black ordering. What is not clear at this time is the rate of convergence of these new methods.

Multigrid. Relaxation methods play a critical role in the multigrid method developed by Brandt [1977], Bank and Sherman [1978], Bank and Dupont [1981], and others. The crucial observation in all multigrid methods is that relaxation methods are very effective at reducing the high frequency component of the error between the computed solution on a particular grid and the true solution, but are very ineffective at reducing the long wave length components. However, if the error is viewed on a sufficiently coarse grid, the long wave length components become high frequency components on that coarser grid, and thus can be effectively reduced by relaxation methods. A very simple multigrid algorithm that incorporates the essential ideas is as follows.

1) Let G^i, $i = 1, \cdots, m$ be a sequence of nested grids covering the domain of interest such that the grid spacing of G^{i-1} is $2h_i$ where h_i is the grid spacing of G^i. G^m is the grid on which the solution is desired and G^1 is an appropriately chosen coarse grid.
2) An approximate solution, u^m, to the discretized differential equation $L^h U^h = F^h$ is obtained by a few iterations of a relaxation method on G^m.
3) For $i = m, \cdots, 2$,
   a) The residual $F^h + L^h u^i \equiv f^i$ is transferred to the grid G^{i-1} by a process known as injection.

    b)   A correction to the solution u^i is computed on the grid G^{i-1}. This step reduces high frequency errors that appear on the coarser grid G^{i-1}.

4)   The solution on the grid G^i is corrected using information interpolated from the grid G^{i-1} for $i = 2, \cdots, m$.

This description, though somewhat simplistic, captures the main steps in the multigrid algorithms; there are many sophisticated variations that are used in practice (see e.g. Hackbusch and Trottenberg [1982]).

The major attraction of the multigrid method is that a wide class of problems discretized on a grid with n points requires only $O(n)$ arithmetic operations to obtain a solution to within the truncation error of the discretization. This attractive computational complexity has led researchers to investigate the method for vector and parallel computers. The relaxation steps on the various grids may be carried out using methods discussed earlier with red-black SOR being the most popular, and the issues raised for effective utilization of the Cray and Cyber 200 remain relevant; however, the nested grids introduce a new difficulty. One would prefer to use vectors of grid points from the finest grid in at least one direction of the region, but then coarser grids require sets of points that are not contiguous in memory and are thus not vectors. For the Cray this means that the vectors must be moved from the vector registers to memory and then the appropriate subset read back into the registers. As indicated earlier, this prevents operation at super vector speeds, and in fact, Caughey [1983] reports performance of 32 MFLOPS on a multigrid code for the solution of the three-dimensional transonic potential flow equations.

For the Cyber 200, hardware instructions for data manipulation must be used to create vectors from the coarse grid points. A detailed description of a multigrid algorithm tailored for the Cyber 205 is given in Barkai and Brandt [1983] and, despite the overhead associated with the data manipulation instructions, the authors report that the vector version of the algorithm runs 15 times faster than the scalar version for a Poisson problem discretized on a 129×129 fine grid. The Cyber 205 is also the target machine for a comparison by Gary, et al. [1983] of multigrid, SOR and a conjugate gradient method preconditioned by a fast Poisson solver. For a three-dimensional diffusion equation with Neumann boundary conditions, they report similar performance for the multigrid and conjugate gradient methods on a $32 \times 32 \times 32$ fine grid. They also conjecture that if the fast Poisson solver based on FFT's were replaced with one based on multigrid then the conjugate gradient algorithm would be 2 to 4 times faster than the pure multigrid algorithm. Finally, Hemker, et al. [1983] discuss the development of fast solvers for elliptic equations based on the multigrid method, including modifications required for the Cyber 205.

Multigrid algorithms have also been considered for parallel arrays. As with vector computers, the nested grids create some difficulties. If these grids are reproduced in their entirety on an array of processors, then in the traditional multigrid formulation most of the processors will be idle most of the time since computation is done only on one grid at a time. On the other hand, if only the finest grid is laid out on the processors, then the pattern of communication between processors will change as the algorithm moves through coarser grids. This can create serious difficulties for a mesh connected array because communication that is between nearest neighbors on the finest grid will not be between nearest neighbors on the coarser grids.

This communication difficulty was recognized by Grosch [1979b] in the first paper that discussed the parallel aspects of multigrid. He compared two strategies. In the first, the finest grid is distributed across the processors and the necessary data is communicated for each grid; in the second, the coarser grid is compressed onto directly connected processors before computation is initiated on that grid. The second strategy is shown to require fewer data transfers. In order to facilitate the communication, Grosch proposed a perfectly shuffled nearest neighbor array by augmenting the nearest neighbor connections with perfect shuffles on each row and column of processors (see Fig. 3.8). A comparison of these two arrays with the optimal paracomputer of Schwartz [1980] indicated that the multigrid algorithm could be implemented with high efficiency on either array.

Brandt [1981] also advocated the perfectly shuffled nearest neighbor array as the basis of a parallel computer for multigrid. In addition he pointed out that the overall efficiency of such computers was going to be adversely affected by processors that were near boundaries or singularities because those processors could have higher computation requirements that would dictate the overall pace of the computation. One suggestion by Brandt that may hold promise for improving the efficiency is to use the finer grid processors to continue relaxation sweeps while the basic algorithm is requiring computation on a coarse grid. A specific example using this idea is given in Brandt [1981] but the results do not indicate that much is accomplished by the extra relaxation sweeps; in fact, if the extra relaxation steps are not done with care, the solution process can deteriorate.

Simultaneous relaxation on all grids is the basis of an algorithm proposed by Gannon and Van Rosendale [1982]. They give a specific way for moving the solutions and residuals between grids and thereby seem to overcome the difficulties encountered by Brandt. This algorithm and two others based on serial implementations of multigrid are analyzed for parallel architectures exhibiting a variety of communication topologies. Simulation experiments based on solving three specific problems indicate that the parallel algorithm would be superior when the computer can efficiently support communication between physically remote processors. One difficulty with the algorithm is that its spectral radius increases with finer grids. This is in sharp contrast to the "classical" multigrid algorithm which owes its attractiveness to the fact that the spectral radius is independent of grid size.

Another study of parallel implementations of multigrid algorithms was done by Chan and Schreiber [1985]. They considered large networks of simple processors with local communication, incorporating the ideas that make systolic architectures attractive for VLSI implementation. Their work contains a careful complexity analysis involving parameters that control key aspects of the multigrid algorithms and a parameter that specifies the number of processors as a function of the number of points in one direction of the grid. By using these complexity results and the concepts of speedup and efficiency they make precise the notion discussed above that if the number of processors is related to the total number of grid points, then many of the processors will be idle a significant amount of the time.

Another approach that has not received study would be to trade off efficiency with performance by sizing the array of processors with one of the coarser subgrids. Processors would go idle only when grids coarser than the one that matched the number of processors were being used. The performance degradation would occur because parallelism available on finer grids could not be utilized due to the restricted number of processors. Using this approach one could be led to different size arrays and different communication topologies depending on the power and cost of the individual processors. It also raises the question of whether one should select a small number of powerful processors or a large number of simple processors for the design of a cost effective computing system for multigrid.

ADI methods. We consider next the Alternating Direction Implicit (ADI) method, which seems, at first glance, to be rather unsatisfactory for parallel computation since it is based on the solution of tridiagonal or small bandwidth systems. Ericksen [1972] and Morice [1972], however, observed that since these tridiagonal systems are independent, they can be solved in parallel. More precisely, we recall that the ADI method—for the model problem and the grid of Fig. 4.1—consists of two half-steps as indicated by the iteration scheme (see, e.g., Varga [1962])

$$(4.6a) \qquad (H + \alpha_k I)x^{k+1/2} = (\alpha_k I - V)x^k + b,$$

$$(4.6b) \qquad (V + \alpha_k I)x^{k+1} = (\alpha_k I - H)x^{k+1/2} + b.$$

The first step, (4.6a), consists of the solution of N tridiagonal systems of size N corresponding to the horizontal lines of the grid, while (4.6b) likewise is the solution of N tridiagonal systems (after permutations of the unknowns) corresponding to the vertical lines. On an array with p processors, p of the tridiagonal systems (4.6a) can be solved in parallel with the usual Gaussian elimination algorithm; it is desirable in this case that N be a multiple of p. On the next half-step, the systems of (4.6b) will be solved in parallel. On a vector computer, the vectors would be aligned across the systems to be solved.

A potential problem on both parallel and vector computers is to arrange the storage so that transfers between half-steps are minimized. This storage problem is particularly pronounced on the Cyber 200 if we vectorize across the tridiagonal systems, and the storage must be rearranged—by the equivalent of a matrix transpose—between each sweep. However, Lambiotte [1975] observed that on the half-sweep that the storage is not correct for the simultaneous solution of the tridiagonal systems, it *is* correct for the solution of the individual tridiagonal systems by the cyclic reduction (CR) method of the previous section. Moreover, the N individual systems may be viewed as forming a single tridiagonal system N times as large and the CR method may be applied to this large system; as we saw in the last section, the larger the system the better. Finally, because the individual systems are uncoupled, the CR method will actually terminate in $\log N$ steps rather than the expected $\log N^2$. Thus, the ADI algorithm is implemented on the Cyber 200 by solving the tridiagonal systems "in parallel" on one half-sweep and as a single large tridiagonal system on the other half-sweep. Lambiotte also discusses a similar strategy for three-dimensional problems.

Block SOR. Many of the same considerations for ADI apply also to the implementation of Successive Line Over-Relaxation (SLOR). If one uses the lexicographic ordering, the same difficulties occur as with point SOR. To circumvent this, Ericksen [1972] and Lambiotte [1975] studied various other orderings, such as a red-back ordering by lines, which allow a number of the tridiagonal systems either to be solved in parallel or as one large tridiagonal system by cyclic reduction. Likewise, multicoloring by lines may be used, if necessary. More recent work on block or line methods has been reported by Boley, et al. [1978], Buzbee, et al. [1979] and Parter and Steuerwalt [1980], [1982], motivated largely by three-dimensional elliptic problems on the Cray-1; see also Faber [1981]. Various possible relaxation schemes using $K \times K$ blocks are discussed, and some of these methods may prove to be attractive. We note also that on arrays, the number of processors may be a key factor in determining the block size. For example, with p processors we would probably try to arrange the computation so that the number of blocks assigned to each processor is a multiple of p.

Somewhat related to block methods, O'Leary and White [1985] consider multisplittings $A = B_i - C_i$, $i = 1, \cdots, k$, of A and an iteration matrix is defined by

$$H = \sum_{i=1}^{k} D_i B_i^{-1} C_i$$

where the D_i are nonnegative diagonal matrices, with $\Sigma D_i = I$. Then the iteration $x^{k+1} = Hx^k + d$ is carried out by executing in parallel the partial iterations defined by $D_i B_i^{-1} C_i$.

Semi-iterative methods. Another group of methods which are potentially useful on vector or parallel machines is semi-iterative (SI) methods. (See, e.g., Young [1971] for a general discussion of these methods.) Consider, for example, the Jacobi–SI method which can be written in the form

(4.7) $$\mathbf{u}^{k+1} = \alpha_k B\mathbf{u}^k + \beta_k \mathbf{u}^k + \gamma_k \mathbf{u}^{k-1}$$

for suitable choice of the parameters α, β, γ. Here B is the Jacobi iteration matrix so that $B\mathbf{u}^k$ is the result of a Jacobi sweep starting from $\mathbf{u}^k$, and the remainder of the calculation of (4.7), once the parameters are known, is ideally suited for vector or parallel machines. Of course, one pays the penalty of additional storage for $\mathbf{u}^{k-1}$. More importantly, the choice of good parameters may be difficult for other than model problems since the optimal parameters are based on a knowledge of the largest and smallest eigenvalues (assumed real) of B. In the case that the coefficient matrix A is symmetric positive definite and has property A, then it is known that the asymptotic rate of convergence of Jacobi–SI is approximately half that of SOR, both using optimal parameters; in this case, Jacobi–SI may not be useful, even on vector computers. However, in more general situations, the rate of convergence of Jacobi–SI may be quite superior to SOR and its somewhat better parallelization properties makes it potentially attractive, provided that reasonable values of the parameters can be chosen. Hayes [1974] and Lambiotte [1975] considered, for the TI-ASC and CDC STAR-100, respectively, the Jacobi–SI method in some detail, as well as other semi-iterative methods such as SSOR-SI and cyclic Chebyshev–SI.

Preconditioned conjugate gradient methods. We turn next to conjugate gradient (CG) methods, which were first developed by Hestenes and Stiefel [1952] as alternatives to Gaussian elimination. Although iterative in nature, they are actually direct methods since, in the absence of rounding error, they converge to the exact solution in no more than n steps for systems of size n. However, in the presence of rounding error, this no longer occurs and the CG methods dropped out of contention as a competitor to elimination. Reid [1971], however, following earlier work of Engeli, et al. [1959], observed that for certain large sparse problems, CG methods gave sufficiently good convergence in far less than n iterations and this spurred revival of these methods for discrete elliptic equations. See, also, Concus, et al. [1976] as one of the important early papers in this revival.

The rate of convergence of the CG methods, viewed as iterative, depends on the condition number, $K(A)$, of A: the smaller the condition number, the more rapid the convergence. Hence, much of the more recent work on CG methods has been devoted to the development of suitable preconditioning strategies. It turns out (see Chandra

[1978] for a full discussion of this and many other basic facts about CG methods) that the preconditioning can be incorporated into the basic CG algorithm as applied to the original matrix A in the following form:

> *Preconditioned Conjugate Gradient* (PCG) *Algorithm*:
> Set $\mathbf{r}^0 = \mathbf{b} - A\mathbf{x}^0$. Solve $M\hat{\mathbf{r}}^0 = \mathbf{r}^0$. Set $\mathbf{p}^0 = \hat{\mathbf{r}}^0$.
> For $k = 0, 1, \cdots$ until convergence
> > Compute $\alpha_k = (\mathbf{r}^k, \mathbf{r}^k)/(\mathbf{p}^k, A\mathbf{p}^k)$, $\mathbf{x}^{k+1} = \mathbf{x}^k + \alpha_k \mathbf{p}^k$
> > Check for convergence. If not, continue
> > Compute $\mathbf{r}^{r+1} = \mathbf{r}^k - \alpha_k A\mathbf{p}^k$. Solve $M\hat{\mathbf{r}}^{k+1} = \mathbf{r}^{k+1}$
> > Compute $\beta_k = (\mathbf{r}^{k+1}, \hat{\mathbf{r}}^{k+1})/(\mathbf{r}^k, \hat{\mathbf{r}}^k)$, $\mathbf{p}^{k+1} = \hat{\mathbf{r}}^{k+1} + \beta_k \mathbf{p}^k$

In the above, $(\mathbf{x}, \mathbf{y})$ denotes the inner product $\mathbf{x}^T\mathbf{y}$ and M arises from the preconditioning of A; if $M = I$, the algorithm reduces to the standard conjugate gradient method. As pointed out earlier, inner products are not attractive computations on parallel or vector computers because of the summation. With this in mind, Van Rosendale [1983b] proposed a modification to the conjugate gradient method that permits the calculation of the inner product from previously computed and stored information. This idea is applicable to the variants discussed below and merits further study.

The matrix M can be viewed as an approximation to A and should satisfy the following criteria (see Concus, et al. [1976]):

a) M is symmetric positive definite.

b) The system $M\hat{\mathbf{r}} = \mathbf{r}$ is "easily" solved.

c) $M^{-1}A$ has "small" or "nearly equal" eigenvalues or has small rank.

Condition c) ensures that the rate of convergence of the PCG method will be faster than that of CG itself. Condition a) is a necessary part of the current theory.

There have been proposed several possible ways to obtain a suitable M; for example:

a) Take M to be the tridiagonal or small bandwidth part of A.

b) Obtain M from an "incomplete" Choleski decomposition of A.

c) Obtain M as a splitting matrix for some suitable iterative method.

As we saw in §3, the solution of small bandwidth systems is not very efficient on vector and parallel computers and option a) has not been very seriously explored for such architectures. We discuss the other two possibilities in more detail.

Probably the most successful preconditioned conjugate gradient methods for sequential computers are the incomplete Choleski conjugate gradient (ICCG) methods (Meijerink and van der Vorst [1977], [1981]). Here, M is obtained as an "incomplete" Choleski decomposition of A; that is, $M = LL^T$ where L is constrained to have the same sparsity pattern as the lower triangular part of A, or some other constraint which makes "easy" both the formation of L and the solution of the linear systems

$$(4.8) \qquad\qquad L\bar{\mathbf{r}} = \mathbf{r}, \qquad L^T\hat{\mathbf{r}} = \bar{\mathbf{r}}$$

with L and L^T as coefficient matrices. The decomposition can be done once and for all at the outset of the iteration and the factor L retained. Thus in the PCG method the solution of $M\hat{\mathbf{r}} = \mathbf{r}$ is effected by the solution of (4.8). As with option a), the solution of these banded systems is not particularly attractive on parallel architectures and most of the research to date has been on different approaches to circumvent this problem.

Rodrigue and Wolitzer [1984a, b], T. Jordan [1982a] and Kershaw [1982] all assume that A is block tridiagonal and that the blocks themselves are tridiagonal, so that

A is a 9-diagonal matrix. Then all of the above authors use some variant of incomplete block cyclic reduction or block odd-even reduction (see §3) to form L and solve the triangular systems (4.8). This extends earlier work of Greenbaum and Rodrigue [1977] who treated 5-diagonal matrices. Kershaw reports that this vectorized algorithm gives speedups of factors of 3 to 6 over the corresponding scalar code on the Cray-1 for certain test problems. Axelsson [1984] also considers block tridiagonal matrices and gives a preconditioning based on an approximate block factorization of A in which the inverses required in a block LU factorization of A are approximated by banded matrices. This is recursive but need be done only once at the beginning of the iteration. The forward and back solves, which are done on each iteration, require only matrix-vector multiplication and are amenable to parallel computation. A modification of the factorization using a cyclic reduction ordering is also presented.

A somewhat different, but related, approach to the ICCG algorithm has been taken by Lichnewsky [1982] (see also Lichnewsky [1983], [1984]) and Schreiber and Tang [1982] by reordering the equations. Lichnewsky, assuming also a block tridiagonal matrix, reorders the blocks in a red-black (odd-even) fashion, and then further reorders within the blocks. The final algorithm is similar to the block cyclic reduction ones described above. Schreiber and Tang use red-black and 4-color reorderings of the equations, and also consider orderings for the ICCG (3) version of Meijerink and van der Vorst [1977], in which L is allowed to have 3 nonzero diagonals outside the nonzero pattern of A. Following Schreiber and Tang's suggestions, Poole and Ortega [1985] give experimental results for two model problems on the Cyber 203 and show that the choice of the multicolor ordering is important in achieving maximum vector lengths. Other related work on ICCG includes Meurant [1985], who develops an incomplete block Choleski decomposition based on work of Concus, et al. [1985], and implements it on the Cray-1, Reiter and Rodrigue [1984], who give an incomplete block Choleski decomposition but based on a permuted form of A, Kowalik and Kumar [1982] who, in the context of a block conjugate gradient algorithm for a multiprocessor environment such as the Denelcor HEP, use a limited Choleski preconditioning scheme in which the diagonal blocks of A are Choleski decomposed, and Jordan and Podsiadlo [1980], who describe a conjugate gradient method implementation on the Finite Element Machine.

Still another approach to ICCG was taken by van der Vorst [1981]. He assumes that A is a 5-diagonal matrix with the main diagonal scaled to 1 and takes L to be the lower triangular part of A so that no Choleski decomposition is really involved. The solution of the systems (4.8) is then effected by a truncated Neumann expansion of $(I-E)^{-1}$, where E is one of the off-diagonals of L.

We now turn to the other general approach to obtaining preconditioning matrices. Let $A = P - Q$ be a splitting of A which defines an iterative method with iteration matrix $G = P^{-1}Q$. If we take m steps of this iterative method towards the solution of the system $A\hat{r} = r$, starting with an initial guess $\hat{r}^0 = 0$, the mth iterate satisfies

$$\hat{r}^{(m)} = (I + G + \cdots + G^{m-1})P^{-1}r$$

so that $\hat{r}^{(m)}$ is the solution of the linear system

(4.9) $$M\hat{r} = r, \qquad M \equiv P(I + \cdots + G^{m-1})^{-1}.$$

As perhaps the simplest example, we can use the Jacobi method in which $P = D$ (the diagonal part of A); the solution of the system (4.9) is then implemented by m Jacobi iterations which, as we have seen, are easily done on parallel and vector architectures.

As a second example, we could use the SOR iteration but this gives a nonsymmetric matrix M. The symmetric SOR (SSOR) iteration (see, e.g. Young [1971]) does lead to a symmetric positive definite M. In the SSOR method, one iterative step consists of an SOR sweep through the grid points followed by a sweep through the grid points in the reverse order. Formally, this can be represented by the splitting $A = P - Q$ with

$$P = \frac{1}{\omega(2-\omega)}(D - \omega L)D^{-1}(D - \omega L^T)$$

where D, $-L$, $-L^T$ are the diagonal, lower and upper triangular parts of A; the solution of the system (4.9) is then implemented by m SSOR iterations in which we would probably use the multicolor orderings previously discussed. We note, however, that the number of iterations of the m-step SSOR PCG method using multicolor orderings to implement the SSOR sweeps may be somewhat more than using the natural ordering. Wang [1982b] has reported on an implementation of a 1-step SSOR PCG using diagonal ordering of the grid points. Although the diagonal orderings do not vectorize as well as the multicolor ones, a reduction in the number of iterations could make them attractive. Rodrigue, et al. [1982] and Lipitakis [1984] have also discussed the use of Jacobi and SSOR preconditioners.

The question arises, in general, as to when the matrix M of (4.9) is symmetric positive definite. Adams [1982], [1985] proved the following, which extended a previous result of Dubois, et al. [1979]. If $P - Q$ is symmetric positive definite with P symmetric and nonsingular, and $G = P^{-1}Q$, then the matrix M of (4.9) is symmetric and

a) For m odd, M is positive definite if and only if P is positive definite.

b) For m even, M is positive definite if and only if $P + Q$ is positive definite.

As an example of the use of this theorem, it can be shown that the matrices P and $P + Q$ of the SSOR splitting are symmetric positive definite if $0 < \omega < 2$. Thus the matrix M for the m-step SSOR preconditioning is symmetric positive definite. Similarly, for the Jacobi iteration, the theorem shows that for m odd, M is positive definite but for m even, M is positive definite only if $D + L + L^T$ is positive definite, which is the classical condition for the convergence of the Jacobi iteration (see, e.g., Young [1971]).

Adams [1982], [1985] has given numerical results for the m-step SSOR PCG method applied to Laplace's equation and a plane stress problem. For these problems the number of iterations required was indeed a decreasing function of m but the point of diminishing returns occurred for $m = 1$ or 2; that is, at least for these problems, it did not pay to use m larger than 2.

Johnson and Paul [1981a, b] and Johnson, et al. [1983] have extended the Dubois, et al. [1979] approach in another direction by replacing the expansion $I + \cdots + G^{m-1}$ in (4.9) by a polynomial in G; that is,

$$(4.10) \qquad M^{-1} = (\alpha_0 I + \alpha_1 G + \cdots + \alpha_{m-1}G^{m-1})P^{-1}.$$

(Actually they considered only the case $P = I$ corresponding to the Jacobi iteration on a matrix assumed to have its main diagonal scaled to be the identity. But Adams [1982], [1985] has shown that the same general approach holds for splittings in which P is symmetric.) The idea now is to choose the parameters α_i in (4.10) so as to ensure the positive definiteness of M and to minimize the ratio of the maximum and minimum eigenvalues of $M^{-1}A$. Johnson, et al. [1983] approach this problem by first noting that, in the case $P = I$, $M^{-1}A$ is also a polynomial, $g(A)$, in A and then choosing the α_i to minimize either $\max g(x)/\min g(x)$ or

$$(4.11) \qquad \int_{\lambda_1}^{\lambda_n} [1 - g(x)]^2 w(x)\, dx.$$

In the above, λ_1 and λ_n are the minimum and maximum eigenvalues of A, w is a suitable weight function, and the max and min of g are taken over the interval $[\lambda_1, \lambda_n]$. Numerical experiments are reported for Laplace's equation on a rectangle with the 5-point star finite difference discretization. They compared their method (with $m = 3$ and the α_i chosen by minimizing (4.11)) with various other methods (ICCG, point SOR, conjugate gradient without preconditioning, etc.) and showed that in every case their method required fewer iterations. Johnson and Lewitt [1982] describe software for implementing the method on the Cyber 205. Saad [1983a, b] has given a rather general analysis of the polynomial preconditioning approach including application to nonsymmetric matrices.

Rodrigue, et al. [1982] consider various versions of the preconditioned conjugate gradient method applied to a diffusion problem on the STAR-100. The diffusion equation is approximated by the method of lines and the corresponding system of ordinary differential equations is solved by an implicit method. It is in carrying out this implicit method that the CG algorithm is used. Preconditioners considered were 1-step Jacobi, 1-step line Jacobi, 1-step Symmetric Gauss–Seidel (SGS) and 1-step SGS with the equations ordered in red/black form. On a particular sample problem run on the STAR-100, the 1-step Jacobi PCG method was the fastest by a good margin.

Saad and Sameh [1981b] and Saad, et al. [1985] also consider the conjugate gradient method as well as a cyclic Chebyshev method and a block Stiefel method treated in an earlier paper (Saad and Sameh [1981a]). Their model problem is a second order elliptic equation with Dirichlet boundary conditions on the unit square, discretized by the 5-point star and with the finite difference equations ordered red-black by lines. They consider the use of a hypothetical array of p processors with a shared memory and report numerical experiments on a sequential computer which showed that the conjugate gradient method was the best of the three, but under certain assumptions on the array they conclude that the block Stiefel method may be superior.

Software for some of the above methods, as well as others, is discussed in several papers by Kincaid (see, e.g., Kincaid, et al. [1984]) and Schonauer (see, e.g. Schonauer, et al. [1983]) and their colleagues.

Variable coefficients. So far we have considered mostly the solution of the linear systems obtained from discretizing a differential equation. With general elliptic operators, additional difficulties will tend to revolve around the best ways to compute and manage storage of the coefficients. For example, consider the equation

$$au_{xx} + bu_{yy} + cu_{zz} = f, \qquad 0 \leq x, y, z \leq 1$$

where a, b, and c are functions of x, y and z. The corresponding difference equations using the usual 7-point formula with $h = \Delta x = \Delta y = \Delta z$ are

$$2\left(a_{i,j,k} + b_{i,j,k} + c_{i,j,k}\right) u_{i,j,k} - a_{i,j,k}\left(u_{i+1,j,k} + u_{i-1,j,k}\right)$$

$$- b_{i,j,k}\left(u_{i,j+1,k} + u_{i,j-1,k}\right) - c_{i,j,k}\left(u_{i,j,k+1} + u_{i,j,k-1}\right) = h^2 f_{i,j,k}.$$

For a sufficiently coarse grid, the coefficients can be computed once and for all and held in five $O(N^3)$ long arrays for vector machines, or distributed over the various processors of an array. But for a moderately fine grid, say $N > 100$, back-up storage may be required and depending upon the complexity of the coefficients, the operating system, and various other factors, it may be more economical to recompute the coefficients at each iteration. This strategy is, of course, common on existing serial machines and the only new factor for vector or parallel computers would be to compute the coefficients in sufficiently large batches—say 1,000-10,000 at a time—so that the computation as well as the subsequent usage in the equation solver can be done efficiently with parallel or vector operations. In particular, recomputation of the coefficients on an array might be useful to save communication time between the processors.

Irregular regions. A less satisfactory situation exists for handling irregular domains. Consider, for example, the grid in Fig. 4.7, where the boundary nodes are indicated by b. One way to handle such a grid is to circumscribe it by a rectangle—the additional grid points thus introduced are indicated in Fig. 4.7 by crosses—and work with the entire rectangular grid. For example, for Laplace's equation and the Jacobi iteration on such a grid, one could use the vector code (4.2) on the Cyber 200 with a control vector, as before, to ensure that the boundary positions are not overwritten. Of course, both additional storage and additional arithmetic are required for the points outside of the domain, and the procedure becomes increasingly less efficient as the domain deviates from rectangular. At some point, it is probably beneficial to use a union of smaller circumscribing rectangles. This, of course, would save considerable storage over a complete circumscribing rectangle but now the rectangles must be processed separately; that is, the code (4.2) must be written separately for each rectangle. Ideally, of course, one would like an ordering of the grid points that would allow processing and storage of only the minimum number of points and still use vectors whose length is the total number of grid points; but such an ordering, if it exists, is not evident and has not appeared in the literature. A more sophisticated approach is to use capacitance matrix methods (see, e.g., O'Leary and Widlund [1979]) but no results for vector computers have been reported.

$$
\begin{array}{cccccccc}
\times & \times & b & b & b & b & \times & \times \\
\times & \times & b & \cdot & \cdot & \cdot & b & \times \\
\times & b & \cdot & \cdot & \cdot & \cdot & \cdot & b \\
b & \cdot & \cdot & \cdot & \cdot & \cdot & \cdot & b \\
\times & b & \cdot & \cdot & \cdot & \cdot & b & \times \\
\times & \times & b & b & b & b & \times & \times \\
\end{array}
$$

FIG. 4.7

Parabolic equations. We turn now to methods for parabolic and hyperbolic equations. As we will see, many of the considerations for time-marching methods on vector and parallel computers are very similar, if not identical, to those for iterative methods for elliptic problems. Explicit methods will tend to be relatively more attractive than on serial computers because of their usually better parallelization properties, but this will not necessarily overcome the stringent stability requirements of small time steps. The question of implicit versus explicit methods, however, is only one part of the broader consideration of how well the method can be adapted to the architectures under consideration. Other aspects which affect this will include the domain, the boundary conditions (and perhaps computational boundary conditions needed for hyperbolic equations and/or higher order methods), the form of the coefficients and whether their calculation can be parallelized, the number of space dimensions, etc.

We will begin with the simple parabolic equation

(4.12)
$$u_t = au_{xx}, \quad t > 0, \quad 0 < x < 1$$

with constant coefficient a and initial-boundary conditions

(4.13)
$$u(0,x) = g(x), \quad u(t,0) = \alpha, \quad u(t,1) = \beta$$

for constant α and β.

Consider first the standard second order Crank–Nicolson scheme

(4.14) $\quad u_j^{k+1} - u_j^k = \dfrac{\mu}{2}\left(u_{j+1}^{k+1} - 2u_j^{k+1} + u_{j-1}^{k+1} + u_{j+1}^k - 2u_j^k + u_{j-1}^k \right), \quad j = 1, \cdots, N$

where $\mu = a\Delta t/(\Delta x)^2$ and u_j^k and u_j^{k+1} indicate values at the current and next time levels, respectively. At each time step, a tridiagonal system of equations must be solved and, as we saw in the last section, this is not particularly efficient on vector or parallel computers with the algorithms now known. By contrast, the simplest explicit method

(4.15)
$$u_j^{k+1} = u_j^k + \mu\left(u_{j+1}^k - 2u_j^k + u_{j-1}^k \right), \quad j = 1, \cdots, N,$$

is mechanistically ideal for vector or parallel computers. Indeed, (4.15) has the same form as the Jacobi iteration applied to a tridiagonal system of equations. On the other hand, the Jacobi iteration could be applied to the tridiagonal systems of the Crank–Nicolson method (4.14). McCulley and Zaher [1974] reported reasonable results with this approach for a diffusion problem on the Illiac IV; in their case 15 Jacobi sweeps sufficed at each time step. More recently, Berger, et al. [1981] discussed a similar approach using the Crank–Nicolson method. With a suitable time step and a suitable predictor formula (forward Euler, Dufort–Frankel) to obtain the initial guess, they found that a single Jacobi sweep on the implicit equations gave sufficient accuracy.

Gelenbe, et al. [1982] also consider the solution of the one-dimensional heat equation. Finite difference discretization is used with a resulting parameterized scheme which includes the fully implicit, fully explicit and Crank–Nicolson methods as special cases. For the implicit schemes, an equation $A\mathbf{u}^{m+1} = B\mathbf{u}^m + \mathbf{c}$ must be solved at each time step and the grid points are ordered in such a way that A has the form

$$\begin{bmatrix} * & * & & & & & & & \\ * & & \cdot & & & & & & \\ & \cdot & \cdot & \cdot & & & & & \\ & & \cdot & \cdot & \cdot & & & & \\ & & & \cdot & \cdot & * & & & \\ & & & & * & * & & & * \\ & & & & & * & * & & \\ & & & & & * & \cdot & \cdot & \cdot \\ & & & & & & \cdot & \cdot & * \\ & & & & * & & & \cdot & * & * \end{bmatrix}$$

That is, it is tridiagonal except for two elements. This sparsity pattern is maintained under LU or Choleski decomposition which is assumed to be done once at the outset. The problem is the forward and back substitutions at each time step. The main purpose of the paper is to give a probabilistic model of the computation on a multiprocessor system and the authors consider in detail the two processor case. The results of their model agree very well with experiments conducted on a system of two LSI 11's at the University of Paris.

The same general considerations apply to problems in two or three space dimensions. For example, for the two-dimensional heat equation, the explicit method corresponding to (4.15) has the same form as the Jacobi iteration. Similarly, the ADI iteration will have the same form as discussed for elliptic equations and the same techniques for handling the tridiagonal systems would again apply.

Hyperbolic equations. For hyperbolic equations, the situation is similar. Indeed, it is somewhat simpler in the sense that except for certain "stiff" systems (i.e. systems with a wide range of eigenfrequencies and characteristic phase velocities), implicit methods are less frequently used, even on serial computers. As a simple example, consider the hyperbolic system

$$(4.16) \qquad \mathbf{u}_t + F(\mathbf{u})_x = 0, \qquad 0 \le x \le 1$$

with suitable initial and boundary conditions. The standard two-step Lax–Wendroff scheme is

$$(4.17) \qquad \begin{aligned} \mathbf{u}_{j+1/2}^{k+1/2} &= \frac{1}{2}\left(\mathbf{u}_{j+1}^k + \mathbf{u}_j^k\right) - \gamma_1\left(F_{j+1}^k - F_j^k\right), \\ \mathbf{u}_j^{k+1} &= \mathbf{u}_j^k - \gamma_2\left(F_{j+1/2}^{k+1/2} - F_{j-1/2}^{k+1/2}\right) \end{aligned}$$

where $\gamma_2 = \Delta t / \Delta x$ and $\gamma_1 = \gamma_2/2$. We see that the two sets of difference equations in (4.17) again have the general form of a Jacobi-like iterative method. A potential difficulty, however, is the evaluation of the vectors derived from $F(\mathbf{u})$. How well this can be done in parallel will depend on the form of F. For example, suppose that $\mathbf{u}$ is the 3-vector of density ρ, momentum m, and energy e and $F = (m, p + m^2/\rho, (e+p)m/\rho)$ where p is given in terms of ρ, and possibly also m and e, by some "equation of state" $p = f(\rho, m, e)$. Then the evaluation of F can be done in parallel as indicated in

$$F_j = \left(m_j, p_j + \frac{m_j^2}{\rho_j}, (e_j + \rho_j)\frac{m_j}{\rho_j}\right), \qquad j = 1, \cdots, N$$

where N is the number of grid points. However, the calculation of the vector of p values may or may not also be computed efficiently by vector operations depending on the form of f. In addition, we will need to handle the given boundary conditions as well as the computational boundary conditions obtained, for example, by extrapolation. Johnson [1984] reviews other considerations in solving the three-dimensional wave equation on vector computers.

Adaptive grids. The preceding discussion has assumed that the grid over which the partial differential equation, be it elliptic, parabolic, or hyperbolic, is discretized remains fixed throughout the solution process. This makes it relatively easy to create vectors out of the grid points or to map grid points onto processors so as to balance computation or to take advantage of the communication topology. On the other hand, many problems or methods may require a dynamically changing grid. It has become common to treat time dependent problems, where some physical phenomena such as a shock wave is moving through the region, with adaptive techiques or grid refinements by adding or repositioning grid points in some area of the region that requires more accuracy. This dynamic change in the grid structure has a serious effect on the data structures for either vector or parallel computers.

For vector computers adaptive computation can be handled in much the same way as described earlier for the multigrid algorithm, that is, by using the data movement instructions on the Cyber 200 or by returning to memory on the Cray. In either case machine efficiency will be reduced. The situation is more difficult for parallel computers. If a subregion assigned to some processor is refined, then either the computation on that processor increases, causing an imbalance, or the region must be redistributed across the array of processors. In either case, an extra burden may be placed on the interprocessor communication mechanism. One approach to the redistribution problem is to have processes that are controlling the computation on a subregion spawn new processes to handle the refined region. Then a computing system that executes these processes on available processors is required. This provides indirect load balancing, but the communication system must be very rich because locality of communication will, in general, be lost.

The FEARS project (see Zave and Rheinboldt [1979] and Zave and Cole [1983]) was an adaptive finite element system which spawned processes to take advantage of parallelism. Refinement was based on a continuous monitoring of the errors and is discussed in detail in Babuska and Rheinboldt [1977]. The subregions, whether refined or not, were organized independently in the spirit of substructuring. As was discussed in §3, this allows for independent parallel computation on the subregions; however, the linear system that connects the subregions was solved sequentially. As reported in Zave and Cole [1983], the sequential solution process requires 70 to 90 percent of the time and thus reduces the overall speedup and efficiency of the process dramatically. Simulations were performed for several computer systems including ZMOB and variations of Cm*. The results of this study are reported in Zave and Cole [1983] and indicate that the majority of the time is spent in communication or waiting.

Adaptive computation in a multigrid setting for three-dimensional problems was the basis of a study by Gannon and Van Rosendale [1984a]. Based on ideas introduced in Van Rosendale [1983a], they also used dynamically spawned processes to provide a framework for the extraction of parallelism. They went on to define an architecture to take advantage of the parallelism. The architecture, which is similar to Cedar (Gajski, et al. [1983]), consists of clusters of processing elements with local memory; the clusters are connected via a crossbar message switching network. Preliminary simulation studies indicate that the system would have a very high level of efficiency due to the fact that over 95 percent of the communication takes place within the clusters.

Spectral methods. A relatively new technique for solving partial differential equations that appears to be appropriate for vector and parallel computers is the spectral method (see e.g. Gottlieb and Orszag [1977] and Voigt, et al. [1984]). In the spectral method, a discrete representation of the solution $u(x)$ of the differential equation $Lu = f$ is approximated by

$$(4.18) \qquad u_n(x) = \sum_{k=0}^{n} a_k \phi_k(x)$$

where the ϕ_k are given functions and (4.18) is evaluated at appropriate points x_j. In order to obtain an approximate solution u_n, expressions for the derivatives of u_n are required based on the form of L. If the x_j and ϕ_k are appropriately chosen, $u_n(x_j)$ and its derivatives may be evaluated using the FFT. Thus, any of the methods alluded to in §3 could be used in a vector or parallel environment. The $u_n(x_j)$ values can be obtained using an appropriate direct or iterative method so again the techniques discussed previously become relevant. Spectral methods are being used extensively on vector computers; see Orszag and Patera [1981a, b], [1983] for the Cray and Bokhari, Hussaini,

Lambiotte and Orszag [1982] for the Cyber 200. These are only representative, and many more references may be found in the bibliography given in Gottlieb, et al. [1984]. There appear to be no studies of the use of spectral methods on parallel computers.

5. Applications. An increasing number of papers have appeared in the last several years describing the use of parallel or vector computers in a variety of application areas. In this section, we will summarize a sampling of this literature without giving extensive details.

Fluid dynamics. The major application area has been fluid dynamics calculations of various kinds. Several early papers described the use of the Illiac IV, some before it was operational. For example, Carroll and Wetherald [1967] discussed the possible application of the Solomon computer—the predecessor, which was never built, of the Illiac IV—to hydrodynamics problems and general circulation weather models in particular; Reilly [1970] considered a Monte Carlo method for the Boltzmann equation; and Ogura, et al. [1972] reviewed the theoretical efficiency of the Illiac IV for hydrodynamics calculations. Wilhelmson [1974] and Ericksen and Wilhelmson [1976] considered convection problems—and in particular the Benard–Rayleigh problem; they used Dufort–Frankel differencing on the diffusion terms, a scheme of Lilly for the convection terms, a fast Fourier method for the Poisson equation, and leap-frog differencing in time. One of the main thrusts of their work was a proper balancing of computation with disk to main memory transfers. Davy and Reinhardt [1975] discussed the application of the Illiac IV to a chemically reacting, inviscid hypersonic flow problem, using MacCormack's method with shock capturing. McCulley and Zaher [1974] reported on the solution of diffusion type equations in a problem that arises in planetary entry.

There were also a number of early papers addressing fluids problems on the TI-ASC and CDC STAR-100. Boris [1976a] applied his flux-corrected transport (FCT) algorithm to continuity type equations on the TI-ASC and concluded that the FCT method is "fully vectorizable". Lambiotte and Howser [1974] compared the ADI method, Brailovskaya's method [1965], and Graves' Partial Implicitization method [1973] on the CDC STAR-100 for the driven cavity problem and concluded that both Brailovskaya's method and Graves' method vectorize well and were the fastest on the STAR even though the ADI method was the fastest on a serial machine. Weilmunster and Howser [1976] considered a boundary layer/shock interaction calculation governed by the full Navier–Stokes equations in two dimensions. They reported speedups of as much as 65 to 1 on the STAR over a corresponding program on a CDC 6600.

Giroux [1977] described the conversion of the HEMP code to the CDC STAR-100. HEMP models two-dimensional deformations, motions and interactions of materials as they are subjected to force fields; it uses an explicit finite difference scheme. Giroux discussed in some detail the many issues in a successful implementation of this procedure on the STAR. The final program showed a speedup of as much as a factor of 5 over the CDC 7600. Soll, et al. [1977] reported on the conversion of the GISS general circulation model to the STAR. Preliminary runs of part of the code showed a speedup of about an order of magnitude over the IBM 360/95.

More recent work has concentrated primarily on the use of the Cyber 200 and Cray series of machines, as well as some parallel arrays. Before describing these developments, we note that there have also been a few recent papers dealing with the older machines. For example, Lomax and Pulliam [1982] (see also Pulliam and Lomax [1979]) report on computations for the unsteady Reynolds-averaged Navier–Stokes equations on the Illiac IV. We also note that there have recently been a number of conference proceedings or anthologies devoted wholly or partly to applications. These include the Los Alamos workshop on vector and parallel computing (Buzbee and Morrison [1978]), applications of the Cray-1 at the Daresbury Laboratory in England (Burke, et al. [1982]), a symposium on applications of the Cray-1 (Cray Research, Inc. [1982]), three symposia on applications of the Cyber 205 (Control Data Corp. [1982], Gary [1984], and Numrich [1985]), and a compilation of articles dealing with a variety of machines but especially the Cray-1 (Rodrigue [1982]). Several of the articles in these sources will be covered in the sequel. We also mention that the book by Gentzsch [1984b] on vectorization contains an entire chapter and an extensive bibliography on applications in fluid dynamics.

Strikwerda [1982] has used the CDC STAR-100 and Cyber 203 for solving the compressible Navier–Stokes equations to obtain laminar flow in converging/diverging nozzles with suction slots. A time-split differencing was used involving three different splittings, one for the parabolic (viscous) terms and two for the hyperbolic (inertial) terms, one for each space direction. Only two-dimensional or axisymmetric problems were handled. The program was coded in SL/1 (Knight and Dunlop [1983]) using 32-bit arithmetic, which was found to give suitable accuracy. For a particular sample calculation for a two-dimensional slotted nozzle, the number of grid points was 12,000 and the number of time steps to convergence was 40,000. The CPU timing for this problem was 1.1×10^{-5} seconds per time step per grid point on the Cyber 203.

Bokhari, Hussaini, Lambiotte and Orszag [1982] treat the Navier–Stokes equations for three-dimensional viscous compressible flow, including compressible shear flows at high Reynolds number, for the Cyber 203 by a mixed spectral/finite difference method, using the one- and two-dimensional FFT codes developed by Korn and Lambiotte [1979] and Lambiotte [1979] as well as techniques for computing derivatives described in Bokhari, Hussaini and Orszag [1982].

Deiwert and Rothmund [1983] use the Cyber 205 for the three-dimensional Navier–Stokes equations modeling boat-tailed afterbodies which are at moderate angles of attack and which contain a centered propulsive jet. There were 216,000 grid points and a database of 5×10^6 words. Fornberg [1983] describes the computation of steady viscous flow past a circular cylinder on the Cyber 205 for Reynolds numbers up to 400. Wu, et al. [1983] report on a direct turbulence simulation which requires solving the time-dependent Navier–Stokes equations in these dimensions. On a two pipeline, 2 million word Cyber 205, they obtain for a $64 \times 64 \times 64$ mesh a computation rate of over 100 MFLOPS using 32-bit arithmetic.

Hankey and Shang [1982] (see also Shang, et al. [1980]) consider three-dimensional Navier–Stokes codes for aerodynamics computations on the Cyber 200 and the Cray-1. Results are given for wind tunnel diffusers, missiles at high angles of attack, self-excited oscillatory flows, etc. Kumar, et al. [1982] report on similar problems for the three-dimensional Navier–Stokes equations on the Cyber 203, including scram-jet inlet and combustor analyses. Rudy [1980] reports on a two-dimensional aerodynamics code for the Cray-1 in which a vectorization of about 85 per cent is attained. This holds the megaflop rate to slightly over 10.

Kascic [1984b] discusses the implementation on the Cyber 205 of a vortex method for the Euler equation for an incompressible inviscid homogeneous fluid. The emphasis is on carefully utilizing the architecture and instruction set of the 205. Woodward [1982] considers various schemes for hydrodynamic problems on different machines and makes a number of worthwhile observations; for example, logical operations are generally slow on vector computers and compress, merge, and mask operations are slow on the Cray-1 since they must be implemented by software. Cox [1983] describes the use of the CDC Cyber 205 on an ocean model. A problem with $18 \times 150 \times 195 = 5 \cdot 10^6$ grid points required approximately 4 seconds per time step, about 4 times faster than the TI-ASC previously used.

Gentzsch [1984a, c] provides an interesting benchmark study that includes a variety of production codes for fluid dynamics problems such as the two-dimensional magneto-hydrodynamic equations, the Navier–Stokes equations, and two- and three-dimensional Euler equations. Results are given for a variety of computers including the Cray-1S, Cyber 205, STAR-100, ICL-DAP, Denelcor HEP and a number of scalar machines. One significant result is that hand coding to improve vectorization improved performance by factors of 2.5 to 5.4 on the Cyber 205 and 1 to 3.3 on the Cray-1S.

Transonic flow. Transonic flow is an important area of aerodynamics which has received considerable attention. Hotovy and Dickson [1979] used a three color ordering of the nodes in connection with a relaxation scheme to solve the two-dimensional small disturbance equation on the CDC STAR-100. They give timing comparisons for this "checkerboard" method on the STAR and an SLOR code on a CDC Cyber 175. On various runs on a 101×41 grid, the STAR was 2.5 to 3 times faster although the checkerboard method required over twice the number of iterations to converge. On a finer (200×80) grid, about 4 times as many iterations were required for the checkerboard method and the STAR was less than twice as fast as SLOR on the Cyber 175.

Redhed, et al. [1979] treated the three-dimensional small disturbance equation of transonic flow on the CDC STAR by using a red-black ordering of grid lines in the cross-flow plane and applying line SOR to all the red columns and then all the black ones. This yields a vector length of half the number of grid points in the cross-flow plane. On a model problem with a relatively coarse mesh, $64 \times 28 \times 20$, they reported a speed-up of a factor of 3.4 over a standard line relaxation code running on a Cyber 175.

Hafez and South [1979] and South, et al. [1980a, b] consider relaxation methods for the full potential equation of transonic flow in both two and three dimensions on the CDC-STAR with comparisons with the CDC 7600 and Cyber 175 as well as the Cray-1. They conclude that point and block SOR using red-black ordering is almost fully vectorizable for this problem. In a subsequent paper, Hafez and Lovell [1983] consider line SOR where m lines are given one color followed by m lines of the other color. They found experimentally the $m = 2$ gives the best results. In earlier work, Keller and Jameson [1978] had used the STAR for the small disturbance equation of transonic flow using a new explicit method. However, they achieved only a speedup of a factor of 1.8 over line overrelaxation running on a Cyber 175.

Melson and Keller [1983] treat the three-dimensional full potential equation in nonconservative form by finite difference methods and in conservative form by finite volumes. Using a test case with a $192 \times 32 \times 32$ grid and a two-color point relaxation scheme (Zebra II, South, et al. [1980b]) they report a computation rate of 26 MFLOPS on the Cyber 203. However, the convergence rate was poor compared with an SLOR algorithm; for related work, see Yu and Rubbert [1982]. Eberhardt, et al. [1984] have studied the mapping of a three-dimensional, implicit, approximate factorization algorithm of the Euler or Navier–Stokes equations onto a two processor Digital Equipment Corp. VAX system that is a reasonable model of a Cray X-MP. They note the importance of careful memory management in a shared memory system and conclude that the algorithm can be implemented on a two processor system with a speedup of 1.9.

Reservoir simulation. There have also been a number of papers dealing with reservoir simulation. Nolen, et al. [1979] give comparisons between the CDC STAR-100 and the Cyber 203 for a model problem. Six different three-dimensional grid sizes were used with the number of unknowns ranging from 2000 to 8000, and tests are reported on the solution of linear equations of these sizes, corresponding to one time step for the time-dependent problem. The algorithms considered for the solution of the linear systems are the D4 method of Gaussian elimination based on the ordering scheme of Price and Coats [1974], and line, 2-line and plane SOR. For the SOR methods, red-black orderings of the grid points are used in such a way that for line SOR the vector length is on the order of $nm/2$, where n and m are the number of grid points in the x and y directions. Similar orderings are used for 2-line and plane SOR, giving smaller vector lengths. Run times on the STAR and 203 are reported for the six grid sizes for the D4 and line-SOR methods; the 203 is somewhat faster on these problems with speedup factors ranging from about 1.15 to 1.5. The only comparison reported for a scalar machine is for Gaussian elimination on a 2000 unknown problem where the 203 was about 14 times faster than a CDC 6600. Additional comparisons between the 203, 205, and Cray-1 are given in Stanat and Nolen [1982]. For the problems reported on, the 205 was a factor of about 2.5 to 3.5 times faster than the 203. The above work and more recent developments are reviewed in Kendall, et al. [1984], which gives comparisons of Cyber and Cray times for various aspects of the algorithms, and also discusses architectural differences that influence implementation decisions.

Also for problems in reservoir simulation, Killough [1979] considered comparisons between the IBM 370/168, with and without an attached IBM 3838 array processor, and the Cray-1. His primary benchmark problem used a three-dimensional rectangular grid with $35 \times 19 \times 5 (=3325)$ grid points and 29 production wells. A production code for the 370/168 was converted and vectorized for the Cray-1 and showed a factor of 12 improvement over the 168. Other papers dealing with reservoir simulation include Buzbee, et al. [1979], Wallis and Grisham [1982] and Kendall, et al. [1983].

Weather prediction. Numerical weather prediction has historically provided one of the major applications of fluid dynamics on high performance computers dating back at least to the ENIAC (see Platzman [1979]). Probably the first weather simulation work with a vector processor was done on the TI-ASC at the Geophysical Fluid Dynamics Laboratory; the mathematical approach and the vectorization of the algorithm are described in the review by Welsh [1982].

The present state of the art for numerical weather prediction is outlined in Cullen [1983]. Current methods for global forecasts use a horizontal grid resolution of 150 km with 15 levels in the vertical direction. Such models require approximately three minutes of Cray or Cyber 200 time for each day of forecast and the average useful forecast period is about four days. Local models for tracking specific meteorological phenomena may have resolution down to 1 km or less. The numerical models tend to be finite differences in the vertical direction and in time and either finite differences or spectral in the horizontal direction. There are at least three other excellent reviews of this field, oriented primarily toward the Cray. Williamson [1983] and Williamson and Swarztrauber [1984] discuss the derivation of the underlying equations, the numerical algorithms, and the implementation on the Cray. Both papers contain extensive bibliographies. Kasahara [1984] reviews many of the decisions that went into the development of computational models and provides some performance figures for the Cray. This paper contains over one hundred references.

Parallel arrays have also been considered for numerical weather prediction. For example, Kopp [1977] and Nagel [1979] report on the use of the SMS201, an array of 128 Intel 8080 microcomputers, for weather problems. They describe a stratified three-dimensional problem in which there are 2000 mesh points at each of three levels and six unknowns per grid point.

Structural analysis. Another major application area for vector and parallel computers is structural analysis. Noor, et al. [1983] discuss the role of high performance computing systems for analysis based on the finite element method concluding that parallelism will play a significant role but suggesting that the full impact will not be reached until software such as programming languages and compilers improves.

Early research on structural analysis applications on vector computers focused on the generation of the elemental stiffness matrix (Noor and Hartley [1978]), and on the solution of the global stiffness matrix system by direct methods (Noor and Fulton [1975], Noor and Voigt [1975], and Noor and Lambiotte [1978]). These studies indicated that the traditional goal in structural analysis of striving for matrices with small bandwidths led to relatively inefficient programs on the STAR-100 because of short vector lengths. The advent of the Cray with its superior performance on short vectors led to renewed interest in the structural analysis community.

A careful study of the effectiveness of the Cray-1 for a structural optimization problem, using an aircraft wing design as motivation, has been done by Venkayya, et al. [1983]. Stresses and displacements are computed and then compared with values representing an acceptable design envelope. Using optimization techniques the process is repeated until satisfactory values are obtained. All modules of the algorithm were studied and those that contributed significantly to the solution time were vectorized. As expected, the most time consuming module was the linear equation solver. The resulting code, which was fine tuned using assembly language, was, on average, 74 times faster than a scalar Fortran code on the Cray for a wing whose discretizations yielded stiffness matrices with 756 to 5280 equations and half-bandwidths of 45 to 105. Another series of applications reported by Goudreau, et al. [1983] involves the study on the Cray-1 of the deformation of large cylindrical cannisters subjected to external loads.

NASTRAN, a large structural analysis program, has been vectorized and is operational on the Cray. The results of this effort for the MacNeal Schwendler Corporation version of the program are reported in Gloudeman and Hodge [1982], Gloudeman [1984] and Gloudeman, et al. [1984]. Timing comparisons with a scalar machine (that unfortunately is not identified) are given. The impact of sparse matrix operations is discussed in McCormick [1982].

Improvements in the Cyber 200 over the STAR-100 have brought about renewed interest in the Cyber 200 for structural analysis. For example, Robinson, et al. [1982] consider implementation of SPAR on the Cyber 203. A different application involves the study of fiber reinforced composite materials that are used in aircraft. At issue is the damage caused by delamination, or the separation of individual layers, in the presence of holes or discontinuities. In particular, Raju and Crews [1982] have conducted a three-dimensional analysis of a four ply laminate with a circular hole, which involved approximately 7000 grid points and a 20 million word database. A very recent study (Raju [1984]) of a more complicated composite led to a system of 100,000 equations with a half-bandwidth of 2700 and a total database of 70 million words. This problem was solved on a two pipeline Cyber 205 with 16 million words of memory at an overall computation rate in excess of 150 MFLOPS using 32-bit arithmetic.

Miscellaneous applications. We next consider a number of miscellaneous application areas. Chang [1982] treats an acoustic wave propagation problem and gives comparisons between a Cray-1S, a Cyber 203, and a Cyber 730. On a series of 6 test problems, the largest of which had vector lengths of 591, the 203 and the Cray were, respectively, 67 to 118 and 142 to 187 times as fast as the 730. The 730, however, had to use disk while the 203 and Cray did not. The state of the art in the dimensional modeling of acoustic phenomena of interest to seismologists is reviewed by Johnson [1984]. This paper contains a discussion of system and programming considerations and gives some performance results. For example, a two-dimensional problem requiring more than ten hours on a Digital Equipment Corp. VAX with an attached Floating Point Systems, Inc. FPS-100 required only eleven minutes on a Cyber 205. Day and Shkoller [1982] describe a three-dimensional code for earthquake analysis which was first developed for the Illiac IV and then converted to a Cray-1. They report that the Cray code ran 75 times faster than its implementation on a UNIVAC 1100/81.

McDonald [1980] used a Chebyshev explicit iteration on an equation of the form $\Delta u + \alpha \cdot \nabla u = f$ with doubly periodic boundary conditions and where α is a function of x and y. This equation arises, for example, in plasma physics. The differential equation is discretized by the usual 5-point star for the Laplacian and centered difference quotients for the first derivatives, and the domain is taken to be a rectangle. Timings from runs on a TI-ASC are given for various grid sizes and compared with an ADI iteration. Although ADI required fewer iterations the superior vectorization properties of the Chebyshev iteration resulted in considerably faster running times.

The design of VLSI devices is another area that is making increased use of high performance computers. This application is reviewed in the article by Fichtner, et al. [1984]. The significant equations are presented and the numerical methods are discussed including implementation considerations for the Cray. The paper includes over eighty references to other literature on integrated circuit design.

Molecular dynamics problems have been solved on the Cray, Cyber and ICL DAP computers. Bowler and Pawley [1984] give a detailed analysis of implementing simulations of phase transitions at the atomic level. They explain how to utilize the architectural features of the DAP and present some representative results. Berendsen, et al. [1984] provide a performance comparison of the Cray-1, the Cyber 203 and 205, the DAP and several scalar computers on some relatively simple molecules. For example, a simple protein in water required approximately 30 hours of Cray or Cyber time. Projections for more complex molecules range up to 10^9 hours of CPU time, making these among the most demanding computational problems.

Other papers dealing with applications include Tennille [1982] on a Cyber 203 code for modeling atmospheric chemical reactions, Liles, et al. [1984] on a thermal-hydraulics program designed to study internal flows in nuclear reactors on the Cray, and Boris and Winsor [1982] on reactive flow problems on the TI-ASC.

There have also been a number of studies of the potential solution of partial differential equations on new, or as yet unbuilt, architectures. As previously mentioned, Dennis and Weng [1977] consider a dataflow architecture for numerical weather predictions. They use as a model the fourth order GISS code (Kalnay-Rivas, et al. [1976]) with a nominal goal of a speedup by a factor of 100 over a 360/95, and describe the computation on a hypothetical dataflow machine. In another study, Dennis [1984a] investigated an implicit algorithm for the solution of the three-dimensional Navier–Stokes equations. The Fortran version of the algorithm was rewritten in Val, and this code was used to outline a hypothetical machine capable of 1000 MFLOPS.

Meyer [1977] studied the possibility of solving the nonlinear Poisson equation $\Delta u = f(u)$ on a hypothetical array of processors. Gallopoulos and McEwan [1983] describe the use of a simulator for the MPP for the solution of the shallow water equations for weather prediction. They conclude that the MPP is suitable for such numerical problems, even though it was designed for image processing. Fox [1984] discusses a variety of applications on a parallel system with the hypercube interconnection and concludes that reasonable efficiency requires the ratio of communication time to computation time be kept near unity.

Epilogue. We have attempted to describe, perhaps too briefly, much of the work which has been done on the use of parallel and vector computers for partial differential equations. Two themes which occurred often, sometimes in conjunction, were decomposition of a problem into independent portions, and reordering of the unknowns in order to enhance such a decomposition. We expect these two themes to be even more prevalent in algorithm development in the future.

It should be clear by now that the differences between vector computers and parallel computers can have a profound effect on the selection of algorithms. In particular, we have shown that computational complexity, the basis for algorithm selection for decades, is still relevant for vector computers because each computation costs some unit of time; however, it is much less relevant for parallel computers for two reasons. First, parallel computers can support extra computation at no extra cost if the computation can be organized properly. Secondly, parallel computers are subject to new overhead costs required, for example, by communication and synchronization that are not reflected by computational complexity. The value of doing extra computation at no extra cost seemed to be recognized by many early researchers in the field who dealt with models consisting of an unbounded number of processors. However, now that parallel systems are available, the research community appears to be focused on analyzing existing algorithms rather than exploring new algorithms for a parallel computing environment. For whatever reason, there have been very few truly new algorithms developed as a result of the opportunities offered by parallelism.

In the near term, it now seems clear that supercomputers from the major vendors will consist of a relatively small number (4, 8, 16, etc.) of powerful vector computers. This is the case with the Cray X-MP, the Cray-2, the Cray-3, the Cyberplus, the ETA GF-10 and the Denelcor HEP. The effective utilization of these machines will require decomposition of the problem into a small number of large, relatively independent parts, and vectorization of the individual parts. The longer term impact of VLSI in the development of highly parallel architectures of thousands of individual processors remains to be seen, although prototypes of such machines are being built. In between these two extremes there are a number of small, new companies offering parallel systems consisting of tens to hundreds of processors each with VAX-like performance. It is simply too early to speculate on how these systems will influence algorithm selection and development.

Thus, all one can say with any certainty is that large scale computing of the future will certainly be highly parallel in one form or another. Even if a single standard parallel system were to exist, there would still be considerable work to be done in the development of efficient numerical algorithms. The likely plethora of different parallel architectures in at least the foreseeable future makes this development more interesting. An especially challenging question is that of software portability across different parallel architectures, a task that will only be feasible when the foundations of parallel computation are much better understood than at present.

BIBLIOGRAPHY

The literature on the parallel solution of partial differential equations is scattered throughout a variety of sources: archival journals in numerical analysis, in computer science, and in a number of engineering disciplines; conference proceedings covering a variety of emphases ranging from numerical analysis to computer architecture to applications such as reservoir simulation; anthologies; and a large number of (as yet) unpublished departmental reports. One of the most important sources is the proceedings of the annual International Conferences on Parallel Processing, listed in the bibliography as Proc. 19xx Int. Conf. Par. Proc. and available through the IEEE Computer Society Press. Certain other conference proceedings and anthologies which have been published in book form we list under the name of the editor (or editors) and then list individual articles with a pointer back to the whole volume; for example, the reference

A. Brandt, [1981], *Multigrid solvers on parallel computers*, in Schultz [1981], pp. 39–83

refers to the article by Brandt in the volume listed under Schultz [1981].

I. Absar, [1983], *Vectorization of a penalty function algorithm for well scheduling*, in Gary [1984], pp. 361–370.

L. Adams, [1982], *Iterative algorithms for large sparse linear systems on parallel computers*, Ph. D. thesis, Univ. Virginia; also published as NASA CR-166027, NASA Langley Research Center, Hampton, VA.

L. Adams, [1983], *An M-step preconditioned conjugate gradient method for parallel computation*, Proc. 1983 Int. Conf. Par. Proc., pp. 36–43.

L. Adams, [1985], *M-step preconditioned conjugate gradient methods*, SIAM J. Sci. Stat. Comput., 6, pp. 452–463.

L. Adams and T. Crockett, [1984], *Modeling algorithm execution time on processor arrays*, Computer, 17, 7, pp. 38–43.

L. Adams and H. Jordan, [1986], Is SOR *color-blind?* ICASE Report No. 84-14, NASA Langley Research Center, Hampton, VA; SIAM J. Sci. Stat. Comput., 7, to appear.

L. Adams and J. Ortega, [1982], *A multi-color SOR method for parallel computation*, Proc. 1982 Int. Conf. Par. Proc., pp. 53–56.

L. Adams and R. Voigt, [1984a], *A methodology for exploiting parallelism in the finite element process*, in Kowalik [1984], pp. 373–392.

L. Adams and R. Voigt, [1984b], *Design, development and use of the finite element machine*, in Parter [1984], pp. 301–321.

T. Agerwala and Arvind, [1982], *Data flow systems*, Computer, 15, 2, pp. 10–13.

H. Ahmed, J. Delosme and M. Mort, [1982], *Highly concurrent computing structures for matrix arithmetic and signal processing*, Computer, 15, 1, pp. 65–82.

G. Amdahl, [1967], *The validity of the single processor approach to achieving large scale computing capabilities*, AFIPS Conf. Proc. 30, pp. 483–485.

G. Anderson and E. Jensen, [1975], *Computer interconnection structures: taxonomy, characteristics, and examples*, ACM Comp. Surveys, 7, pp. 197–213.

C. Arnold, [1982], *Performance evaluation of three automatic vectorizer packages*, Proc. 1982, Int. Conf. Par. Proc., pp. 235–242.

C. Arnold, [1983], *Vector optimization on the CYBER 205*, Proc. 1983 Int. Conf. Par. Proc., pp. 530–536.

C. Arnold, [1984], *Machine independent techniques for scientific supercomputing*, Proc. COMPCON 84, IEEE Comp. Sci. Conf., pp. 74–83.

C. Arnold, M. Parr, and M. Dewe, [1983], *An efficient parallel algorithm for the solution of large sparse linear matrix equations*, IEEE Trans. Comput., C-32, pp. 265–273.

Arvind and R. Bryant, [1979], *Parallel computers for partial differential equations simulation*, Proc. Scientific Computer Information Exchange Meeting, Livermore, CA, 1979, pp. 94–102.

Arvind and V. Kathail, [1981], *A multiple processor data flow machine that supports generalized procedures*, 8th Annual Symp. Comp. Arch., May, pp. 291–302.

S. Arya and D. Calahan, [1981], *Optimal scheduling of assembly language kernels for vector processors*, 19th Allerton Conf. on Comm. Control and Computers, Univ. Illinois, Champaign–Urbana.

S. Askew and F. Walkden, [1984], *On the design and implementation of a package for solving a class of partial differential equations*, in Paddon [1984], pp. 107–114.

A. Avizienis, M. Evcegovac, T. Lang, P. Sylvain and A. Thomasian, [1977], *An investigation of fault-tolerant architectures for large scale numerical computing*, in Kuck, et al. [1977], pp. 159–183.

O. AXELSSON, [1984], *A survey of vectorizable preconditioning methods for large scale finite element matrix problems*, Center for Numerical Analysis Report No. CNA-190, Univ. Texas, Austin.

I. BABUSKA AND W. RHEINBOLDT, [1977], *Computational Aspects of Finite Element Analysis*, Mathematical Software III, J. Rice, ed., Academic Press, New York, pp. 223–253.

J. BACKUS, [1978], *Can programming be liberated from the von Neumann style? A functional style and its algebra of programs*, Comm. ACM, 21, pp. 613–641.

J.-L. BAER, [1980], *Supercomputers*, in Computer Systems Architecture, Computer Science Press, Los Alamitos, CA.

J.-L. BAER, [1984], *Computer architecture*, Computer, 17, 10, pp. 77–87.

W. BALLHAUS, [1984], *Computational aerodynamics and supercomputers*, Proc. COMPCON 84, IEEE Comp. Soc. Conf., pp. 3–14.

R. BANK AND T. DUPONT, [1981], *An optimal order process for solving elliptic finite element equations*, Math. Comp., 36, pp. 35–51.

R. BANK AND A. SHERMAN, [1978], *Algorithmic aspects of the multi-level solution of finite element equations*, Center for Numerical Analysis Report No. CNA-144, Univ. Texas, Austin.

D. BARKAI AND A. BRANDT, [1983], *Vectorized multigrid Poisson solver for the CDC Cyber 205*, Appl. Math. Comp., 13, pp. 215–228.

D. BARKAI, K. MORIARTY AND C. REBBI, [1984a], *A highly optimized vectorized code for Monte Carlo simulation of SU(3) lattice gauge theories*, Comp. Phys. Comm., 32, pp. 1–9.

D. BARKAI, K. MORIARTY AND C. REBBI, [1984b], *A highly optimized vectorized code for Monte Carlo simulation of SU(3) lattice gauge theories*, Proc. 1984 Int. Conf. Par. Proc., pp. 101–108.

D. BARKAI, K. MORIARTY AND C. REBBI, [1984c], *A modified conjugate gradient solver for very large systems*, in Numrich [1985].

J. BARLOW AND I. IPSEN, [1984], *Parallel scaled Givens rotations for the solution of linear least squares problems*, Dept. Computer Science Report YALEU/DCS/RR-310, Yale Univ., New Haven, CT.

R. BARLOW AND D. EVANS, [1982], *Synchronous and asynchronous iterative parallel algorithms for linear systems*, Comput. J., 25, pp. 56–60.

R. BARLOW, D. EVANS AND J. SHANEHCHI, [1984], *Sparse matrix vector multipliction on the DAP*, in Paddon [1984], pp. 147–155.

G. BARNES, R. BROWN, M. KATZ, D. KUCK, D. SLOTNICK AND R. STOKER, [1968], *The Illiac IV computer*, IEEE Trans. Comput., C-17, pp. 746–757.

K. BATCHER, [1974], STARAN *parallel processor system hardware*, AFIPS Conf. Proc. 43, NCC, pp. 405–410.

K. BATCHER, [1979], MPP—*A massively parallel processor*, Proc. 1979 Int. Conf. Par. Proc., p. 249.

K. BATCHER, [1980], *Design of a massively parallel processor*, IEEE Trans. Comput., C-29, pp. 836–840.

G. BAUDET, [1977], *Iterative methods for asynchronous multiprocessors*, in Kuck et al. [1977], pp. 309–310.

G. BAUDET, [1978], *Asynchronous iterative methods for multiprocessors*, J. ACM, 25, pp. 226–244.

G. BEHIE AND P. FORSYTH, [1984], *Incomplete factorization methods for fully implicit simulation of enhanced oil recovery*, SIAM J. Sci. Stat. Comput., 5, pp. 543–561.

V. BENES, [1962], *Heuristic remarks and mathematical problems regarding the theory of connecting systems*, Bell Syst. Tech. J., 41, pp. 1201–1247.

V. BENES, [1965], *Mathematical Theory of Connecting Networks and Telephone Traffic*, Academic Press, New York.

H. BERENDSEN, W. VAN GUNSTEREN, AND J. POSTMA, [1984], *Molecular dynamics on* CRAY, CYBER *and* DAP, in Kowalik [1984], pp. 425–438.

M. BERGER, J. OLIGER AND G. RODRIGUE, [1981], *Predictor–corrector methods for the solution of time dependent parabolic problems on parallel processors*, in Schultz [1981], pp. 197–202.

P. BERGER, P. BROUAYE AND J. SYRE, [1982], *A mesh coloring method for efficient* MIMD *processing in finite element problems*, Proc. 1982 Int. Conf. Par. Proc., pp. 41–46.

M. BERZINS, T. BUCKLEY AND P. DEW, [1984], *Path Pascal simulation of multiprocessor lattice architectures for numerical computations*, in Paddon [1984], pp. 25–33.

V. BHAVSAR AND U. GUJAR, [1984], VLSI *algorithms for Monte Carlo solutions of partial differential equations*, in Vichnevetsky and Stepleman [1984], pp. 268–276.

V. BHAVSAR AND J. ISAAC, [1982], *Design and analysis of parallel algorithms for Monte Carlo techniques*, Proc. 10th IMACS World Congress on Systems Simulation and Scientific Computation, vol. 1, IMACS, New Brunswick, NJ, pp. 323–325.

L. BHUYAN AND D. AGRAWAL, [1984], *Generalized hybercube and hyberbus structures for a computer network*, IEEE Trans. Comput., 33, pp. 323–333.

D. BINI, [1984], *Parallel solution of certain Toeplitz linear systems*, SIAM J. Comput., 13, pp. 368–476.

S. BIRINGEN, [1983a], *A numerical simulation of transition in plane channel flow*, AIAA Paper No. 83-47, January, Reno, NV.

S. BIRINGEN, [1983b], *Simulation of late transition in plane channel flow*, Proc. Third International Conference on Numerical Methods in Laminar and Turbulent Flow, August, Seattle, WA.

G. BIRKHOFF AND A. SCHOENSTADT, eds., [1984], *Elliptic Problem Solvers*, Academic Press, New York.

J. BISKEBORN, [1983], *A multiprocessor implementation of* CSP, Computer Science Dept. Report, Univ. Colorado, Boulder.

J. BOISSEAU, M. ENSELME, D. GUINRAUD AND P. LEED, [1982], *Potential assessment of a parallel structure for the solution of partial differential equations*, Rech. Aerosp.

A. BOJANCZYK, R. BRENT AND H. KUNG, [1984], *Numerically stable solution of dense systems of linear equations using mesh-connected processors*, SIAM J. Sci. Stat. Comput. 5, pp. 95–104.

S. BOKHARI, [1979], *On the mapping problem*, Proc. 1979 Int. Conf. Par. Proc., pp. 239–248.

S. BOKHARI, [1981], *On the mapping problem*, IEEE Trans. Comput., C-30, pp. 207–214.

S. BOKHARI, [1984], *Finding a maximum on an array processor with a global bus*, IEEE Trans. Comput., C-33, pp. 133–139.

S. BOKHARI, M. HUSSAINI, J. LAMBIOTTE AND S. ORSZAG, [1982], *Navier–Stokes solution on the* CYBER-203 *by a pseudospectral technique*, Second IMACS International Symposium on Parallel Computation, Nov. 9–11, 1982, Newark, DE, pp. 305–307.

S. BOKHARI, M. HUSSAINI AND S. ORSZAG, [1982], *Fast orthogonal derivatives on the* STAR, Comput. Math. Appl., 8, pp. 367–377.

D. BOLEY, [1978], *Vectorization of block relaxation techniques: some numerical experiments*, Proc. 1978 LASL Workshop on Vector and Parallel Processors, Los Alamos, NM.

D. BOLEY, B. BUZBEE AND S. PARTER, [1978], *On block relaxation techniques*, Mathematics Research Center Report 1860, Univ. Wisconsin, Madison.

L. BONEY AND R. SMITH, [1979], *A vectorization of the Hess–McDonnel–Douglas potential flow program NUED for the STAR-100 computer*, NASA TM-78816, NASA Langley Research Center, Hampton, VA.

D. BOOK, ed., [1981], *Finite Difference Techniques for Vectorized Fluid Dynamics Calculations*, Springer-Verlag, New York.

J. BORIS, [1976a], *Flux-corrected transport modules for solving generalized continuity equations*, Naval Research Laboratory Report 3237, Washington, DC.

J. BORIS, [1976b], *Vectorized tridiagonal solvers*, Naval Research Laboratory Report 3048, Washington, DC.

J. BORIS AND N. WINSOR, [1982], *Vectorized computation of reactive flow*, in Rodrigue [1982], pp. 173–215.

A. BOSSAVIT, [1982], *On the vectorization of algorithms in linear algebra*, Proc. 10th IMACS World Congress on Systems Simulation and Scientific Computation, vol. 1, IMACS, New Brunswick, NJ, pp. 95–97.

W. BOUKNIGHT, S. DESENBERG, D. MCINTYRE, J. RANDALL, A. SAMEH AND D. SLOTNICK, [1972], *The Illiac IV system*, Proc. IEEE, 60, pp. 369–379.

K. BOWLER AND G. PAWLEY, [1984], *Molecular dynamics and Monte Carlo simulations in solid state and elementary particle physics*, Proc. IEEE, 72, pp. 42–55.

P. BRADLEY, D. DWOYER AND J. SOUTH, [1984], *Vectorized schemes for conical flow using the artificial density method*, AIAA Paper 84-1062, January.

P. BRADLEY, P. SIEMERS AND K. WEILMUENSTER, [1982], *Comparison of shuttle flight pressure data to computational and wind-tunnel results*, J. Spacecraft and Rockets, 19, pp. 419–422.

I. BRAILOVSKAYA, [1965], *A difference scheme for numerical solution of the two-dimensional non-stationary Navier–Stokes equations for a compressible gas*, Soviet Physics Doklady, 10, pp. 107–110.

A. BRANDT, [1977], *Multigrid adaptive solutions to boundary value problems*, Math. Comp., 31, pp. 333–390.

A. BRANDT, [1981], *Multigrad solvers on parallel computers*, in Schultz [1981], pp. 39–83.

R. BRENT AND F. LUK, [1983a], *Computing the Cholesky factorization using a systolic architecture*, Proc. 6th Australian Computer Science Conf., pp. 295–302.

R. BRENT AND F. LUK, [1983b], *A systolic array for the linear time solution of Toeplitz systems of equations*, J. VLSI and Computer Systems, 1, pp. 1–22.

B. BRODE, [1981], *Precompilation of Fortran programs to facilitate array processing*, Computer, 14, 9, pp. 46–51.

G. BROOMELL AND J. HEATH, [1983], *Classification categories and historical development of circuit switching topologies*, Comp. Surveys, 15, pp. 95–134.

J. BROWNE, [1984a], *Parallel architecture for computer systems*, Physics Today, 37, 5, pp. 28–35.

J. BROWNE, [1984b], *TRAC: An environment for parallel computing*, Proc. COMPCON 84, IEEE Comp. Soc. Conf., pp. 294–299.

I. BUCHER AND T. JORDAN, [1984a], *Linear algebra programs for use on a vector computer with a secondary solid state storage device*, in Vichnevetsky and Stepleman [1984], pp. 546–550.

I. BUCHER AND T. JORDAN, [1984b], *Solving very large elliptic problems on a supercomputer with solid state disk*, J. Comp. Phys., 55, pp. 340–345.

O. BUNEMAN, [1969], *A compact non-iterative Poisson solver*, Institute for Plasma Research Report. 294, Stanford Univ., Stanford, CA.

P. BUNING AND J. LEVY, [1979], *Vectorization of implicit Navier–Stokes codes on the* CRAY-1 *computer*, Dept. Aeronautics and Astronautics., Stanford Univ., Stanford, CA.

P. BURKE, B. DAVIES AND D. EDWARDS, eds., [1982], *Some research applications on* the CRAY-1 *computer at the Daresbury Laboratory*, 1979-81, Daresbury Laboratory, England.

P. BURKE AND L. DELNES, eds., [1982], *Proceedings of the International Conference on Vector and Parallel Processors in Computational Science*, Chester, England, August, 1981; Comp. Phys. Comm., 26 (1982), pp. 217–488.

Burroughs Corp., [1979], NAS *facility feasibility study*, Final Report, Contract No. NAS2-9897.

T. BUTLER, J. CLOUTMAN AND J. RAMSHAW, [1981], *Multidimensional numerical simulation of reactive flow in internal combustion engines*, Prog. Energy Combust. Sci., 7, pp. 293–315.

B. BUZBEE, [1973], *A fast Poisson solver amenable to parallel computation*, IEEE Trans. Comput., C-22, pp. 793–796.

B. BUZBEE, [1981], *Implementing techniques for elliptic problems on vector processors*, in Schultz [1981], pp. 85–98.

B. BUZBEE, [1983a], *Vectorization of algorithms for solving systems of elliptic difference equations*, in Noor [1983], pp. 81–88.

B. BUZBEE, [1983b], *Two parallel formulations of particle-in-cell models*, Los Alamos National Laboratory Report LA-UR-83-413, Los Alamos, NM.

B. BUZBEE, [1983c], *Remarks for the IFIP congress '83 panel on how to obtain high performance for high-speed processors*, Los Alamos National Laboratory Report LA-UR-83-1392, Los Alamos, NM.

B. BUZBEE, [1984a], *Gaining insight from supercomputing*, Proc. IEEE, 72, pp. 19–21.

B. BUZBEE, [1984b], *Application of* MIMD *machines*, Los Alamos National Laboratory Report LA-UR-84-2004, Los Alamos, NM.

B. BUZBEE, D. BOLEY AND S. PARTER, [1979], *Applications of block relaxation*, Proc. 1979 AIME Fifth Symposium on Reservoir Simulation.

B. BUZBEE, R. EWALD, AND J. WORLTON, [1982], *Japanese supercomputer technology*, Science, 218, 4578, pp. 1189–1193.

B. BUZBEE, G. GOLUB AND C. NIELSON, [1970]. *On direct methods for solving Poisson's equation*, SIAM J. Numer. Anal., 7, pp. 627–656.

B. BUZBEE, G. GOLUB AND J. HOWELL, [1977], *Vectorizations for the* CRAY-1 *of some methods for solving elliptic difference equations*, in Kuck, et al. [1977], pp. 255–271.

B. BUZBEE AND J. MORRISON, eds., [1978], *Proc.* 1978 LASL *Workshop on Vector and Parallel Processors*, Los Alamos, NM.

B. BUZBEE AND D. SHARP, [1985], *Perspectives on supercomputing*, Science, 227, pp. 591–597.

B. BUZBEE, J. WORLTON, G. MICHAEL AND G. RODRIGUE, [1980], DOE *research in utilization of high performance systems*, Los Alamos National Laboratory Report LA-8609-MS, Los Alamos, NM.

D. CALAHAN, [1973], *Parallel solution of sparse simultaneous linear equations*, Proc. 11th Allerton Conference on Circuit and System Theory, Univ. Illinois, Urbana, pp. 729–738.

D. CALAHAN, [1975], *Complexity of vectorized solution of two-dimensional finite element grids*, Systems Engineering Laboratory Report 91, Univ. Michigan, Ann Arbor.

D. CALAHAN, [1979a], *A block-oriented sparse equation solver for the* CRAY-1, Proc. 1979 Int. Conf. Par. Proc., pp. 116–123.

D. CALAHAN, [1979b], *Vectorized sparse elimination*, Proc. Sci. Computer Information Exchange Meeting, Livermore, CA.

D. CALAHAN, [1980], *Multi-level vectorized sparse solution of* LSI *circuits*, Proc. IEEE Conf. on Circuits and Computers, Rye, NY, October, pp. 976–979.

D. CALAHAN, [1981a], *Direct solution of linear equations on the* CRAY-1, CRAY Channels, 3, pp. 1–5.

D. CALAHAN, [1981b], *Performance of linear algebra codes on the* CRAY-1, SPE J., pp. 558–564.

D. CALAHAN [1981c], *Sparse vectorized direct solution of elliptic problems*, in Schultz [1981], pp. 241–245.

D. CALAHAN, [1982a], *High performance banded and profile equation solvers for the* CRAY-1: *The unsymmetric case*, Systems Engineering Laboratory Report 160, Univ. Michigan, Ann Arbor.

D. CALAHAN, [1982b], *Vectorized direct solvers of* 2-D *grids*, Proc. 6th Symp. Reservoir Simulation, pp. 489–506.

D. CALAHAN, [1983], *Tasking studies in solving a linear algebra problem on a* CRAY-*class multiprocessor*, Supercomputer Algorithm Research Laboratory Report No. SARL 2, Univ. Michigan, Ann Arbor.

D. CALAHAN, [1984], *Influence of task granularity on vector multiprocessor performance*, Proc. 1984 Int. Conf. Par. Proc., pp. 278–284.

D. CALAHAN AND W. AMES, [1979], *Vector processors: models and application*, IEEE Trans. Circuits and Syst., CAS-26, pp. 715–776.

D. CALAHAN, W. AMES AND E. SESEK, [1979], *A collection of equation solving codes for the* CRAY-1, Systems Engineering Laboratory Report, Univ. Michigan, Ann Arbor.

D. CALAHAN, W. JOY AND P. ORBITS, [1976], *Preliminary report on results of matrix benchmarks on vector processors*, Systems Engineering Laboratory Report, Univ. Michigan, Ann Arbor.

G. CAREY, [1981], *High speed processors and implications for algorithms and methods*, in Nonlinear Finite Element Analysis—Structural Mechanics, W. Wunderlich, E. Stein and K. Bathe, eds., Springer-Verlag, Berlin.

A. CARROLL AND R. WETHERALD, [1967], *Application of parallel processing to numerical weather prediction*, J. ACM, 14, pp. 591–614.

D. CASASENT, [1984], *Acoustooptic linear algebra processors—architectures, algorithms and applications*, Proc. IEEE, 72, pp. 831–849.

D. CAUGHEY, [1983], *Multigrid calculations of three-dimensional transonic potential flows*, Appl. Math. Comp. 13, pp. 241–260.

D. CAUGHEY, P. NEWMAN AND A. JAMESON, [1978], *Recent experiences with three dimensional transonic potential flow calculations*, NASA TM 78733, NASA Langley Research Center, Hampton, VA.

T. CHAN AND Y. SAAD, [1985], *Multigrid algorithms on the hypercube multiprocessor*, Research Report YALEU/DCS/RR-368, Yale Univ., New Haven, CT.

T. CHAN AND R. SCHREIBER, [1985], *Parallel networks for multigrid algorithms: architecture and complexity*, Computer Science Report 262, Yale Univ., New Haven, CT, SIAM J. Sci. Stat. Comput., 6 (1985), pp. 698–711.

R. CHANDRA, [1978], *Conjugate gradient methods for partial differential equations*, Ph. D. Thesis, Dept. Computer Science, Yale Univ., New Haven, CT.

S. CHANG, [1982], *Borehole acoustic simulation on vector computers*, in Control Data Corp., [1982].

D. CHAPMAN, [1979], *Computational aerodynamics development and outlook*, 17th Aerospace Sciences Meeting, AIAA paper 79-0129.

D. CHAZAN AND W. MIRANKER, [1969], *Chaotic relaxation*, J. Lin. Alg. Appl., 2, p. 199–222.

A. CHEN AND C. WU, [1984], *Optimum solution to dense linear systems of equations*, Proc. 1984 Int. Conf. Par. Proc., pp. 417–424.

S. CHEN, [1982], *Polynomial scaling in the conjugate gradient method and related topics in matrix scaling*, Ph. D. dissertation, Dept. Computer Science, Pennsylvania State Univ., University Park.

S. CHEN, [1975], *Speedup of iterative programs in multi-processing systems*, Ph. D. Thesis, Dept. Computer Science, Univ. Illinois, Champaign-Urbana.

S. CHEN, [1984], *Large-scale and high-speed multiprocessor system for scientific applications:* CRAY X-MP-2 *series*, in Kowalik [1984], pp. 59–67.

S. CHEN, J. DONGARRA AND C. HSUING, [1984], *Multiprocessing linear algebra algorithms on the* CRAY X-MP-2: *Experiences with small granularity*, J. Par. and Dist. Comp., 1, pp. 22–31.

S. CHEN AND D. KUCK, [1975], *Time and parallel processor bounds for linear recurrence systems*, IEEE Trans. Comput., C-24, pp. 101–117.

S. CHEN, D. KUCK AND A. SAMEH, [1978], *Practical parallel and triangular systems solvers*, ACM Trans. Math. Software 4, pp. 270–77.

S. CHEN AND A. SAMEH, [1975], *On parallel triangular system solvers*, Proc. 1975 Sagamore Conf. Par. Proc., pp. 237–38.

M. CHERN AND T. MURATA, [1983a], *A fast algorithm for concurrent LU decomposition and matrix inversion*, Proc. 1983 Int. Conf. Par. Proc., pp. 79–86.

M. CHERN AND T. MURATA, [1983b], *Efficient matrix multiplication on a concurrent data-loading array processor*, Proc. 1983 Int. Conf. Par. Proc., pp. 90–94.

R. CHIMA AND G. JOHNSON, [1983], *Efficient solution of the Euler and Navier–Stokes equations with a vectorized multiple-grid algorithm*, AIAA Paper 83-1893.

N. CHRIST AND A. TERRANO, [1984], *A very fast parallel processor*, IEEE Trans. Comput., 33, pp. 344–350.

C. CLOS, [1953], *A study of non-blocking switching networks*, Bell Syst. Tech. J., 32, pp. 406–424.

J. COCKE AND D. SLOTNICK, [1958], *The use of parallelism in numerical calculations*, IBM Research Memorandum RC-55, IBM Res. Center,

W. COLLIER, C. MCCALLIEN AND J. ENDERBY, [1984], *Tough problems in reactor design*, in Paddon [1984], pp. 91–106.

P. CONCUS, G. GOLUB AND G. MEURANT, [1985], *Block preconditioning for the conjugate gradient method*, SIAM J. Sci. Stat. Comput., 6, pp. 220–252.

P. CONCUS, G. GOLUB AND D. O'LEARY, [1976], *A generalized conjugate gradient method for the numerical solution of elliptic partial differential equations*, in Sparse Matrix Computations, J. Bunch and D. Rose, eds., Academic Press, New York, pp. 309–322.

V. CONRAD AND Y. WALLACH, [1977], *Iterative solution of linear equations on a parallel processor system*, IEEE Trans. Comput., C-26, pp. 838–847.

Control Data Corporation, [1979], *Final report. Feasibility study for NASF*, NASA Contractor Report NAS2-9896.

Control Data Corporation, [1982], *Proc. Symposium CYBER 205 Applications*, Ft. Collins, CO.

M. COX, [1983], *Ocean modeling on the Cyber 205 at GFDL*, in Gary [1984], pp. 27–32.

Cray Research, Inc., [1982], *Science, engineering and the CRAY-1*. Proceedings of a Cray Research Inc. Symposium.

L. CSANKY, [1976], *Fast parallel matrix inversion algorithms*, SIAM J. Comput., 5, 618–623.

M. CULLEN, [1983], *Current progress and prospects in numerical techniques for weather prediction models*, J. Comp. Phys., 50, pp. 1–37.

W. CYRE, C. DAVIS, A. FRANK, L. JEDYNAK, M. REDMOND AND V. RIDEOUT, [1977], *WISPAC: a parallel array computer for large scale system simulation*, Simulation, 11, pp. 165–172.

W. DAVY AND W. REINHARDT, [1975], *Computation of shuttle non-equilibrium flow fields on a parallel processor*, NASA SP-347, NASA Ames Research Center, Moffett Field, CA, pp. 1351–1376.

S. DAY AND B. SHKOLLER, [1982], *A 3-D earthquake model*, in Control Data Corp., [1982].

G. DEIWERT AND H. ROTHMUND, [1983], *Three dimensional flow over a conical afterbody containing a centered propulsive jet: a numerical simulation*, AIAA 16th Fluid and Plasma Dynamics Conference; also in Gary [1984], pp. 187–200.

J.-M. DELOSME AND I. IPSEN, [1984], *Efficient parallel solution of linear systems with hyperbolic rotations*, Dept. Computer Science Report YALEU/CSD/RR-341, Yale Univ., New Haven, CT.

J.-M. DELOSME AND M. MORF, [1981], *Scattering arrays for matrix computations*, SPIE 25th Tech. Symp., San Diego, CA.

P. DELSARTE, Y. GENIN AND Y. KAMP, [1980], *A method of matrix inverse triangular decomposition based on contiguous principle submatrices*, J. Lin. Alg. Appl., 31, pp. 194–212.

L. DELVES, A. SAMBA AND J. HENDRY, [1984], *Band matrices on the DAP*, in Paddon [1984], pp. 167–183.

B. DEMBAST AND K. NEVES, [1977], *Sparse triangular factorization on vector computers*, in Exploring Applications of Parallel Processing to Power System Analysis, Electric Power Res. Inst. Rep. EE 566-SR.

J. DEMINET, [1982], *Experience with multiprocessor algorithms*, IEEE Trans. Comput., C-31, pp. 278–288.

J. DENNIS, [1980], *Data flow supercomputers*, Computer, 13, 11, pp. 48–56.

J. DENNIS, [1982], *High speed data flow computer architecture for the solution of the Navier–Stokes equations*, Massachusetts Institute of Technology Laboratory for Computer Science Report, Cambridge, MA.

J. DENNIS, [1984a], *High speed data flow computer architecture for the solution of the Navier–Stokes equations*, Computation Structures Group Memo 225, Massachusetts Institute of Technology Laboratory for Computer Science, Cambridge, MA.

J. DENNIS, [1984b], *Data flow ideas for supercomputers*, Proc. COMPCON 84, IEEE Comp. Soc. Conf., pp. 15–20.

J. DENNIS, G. R. GAO AND K. TODD, [1984], *Modeling the weather with a dataflow supercomputer*, IEEE Trans. Comp., C-33, pp. 592–603.

J. DENNIS AND K. WENG, [1977], *Application of data flow computation to the weather problem*, in Kuck, et al. [1977], pp. 143–157.

J. DEUTSCH AND H. NEWTON, [1984], *MSPLICE: A multiprocessor based circuit simulator*, Proc. 1984 Int. Conf. Par. Proc., pp. 207–214.

M. DIAMOND, [1975], *The stability of a parallel algorithm for the solution of tridiagonal linear systems*, Proc. 1975 Sagamore Conf. Par. Proc., p. 235.

D. DODSON, [1981], *Preliminary timing study for the CRAYPACK Library*, Boeing Computer Services Internal Memorandum G4550-CM-39, Seattle, WA.

D. DODSON AND J. LEWIS, [1982], *Improving the performance of a sparse matrix solver on the CRAY-1*, in CRAY Research, Inc. [1982], pp. 13–15.

J. DONGARRA, [1978], *Some Linpack timings on the CRAY-1*, Proc. 1978 LASL Workshop on Vector and Parallel Processors, pp. 58–75.

J. DONGARRA, [1983], *Redesigning linear algebra algorithms*, EDF Bulletin de la Direction des Études et Recherches, Ser. C, 1, pp. 51–59.

J. DONGARRA, [1984], *Performance of various computers using standard linear equations software in a Fortran environment*, Argonne National Laboratory Report MCA-TM-23, Argonne, IL.

J. DONGARRA AND S. EISENSTAT, [1984], *Squeezing the most out of an algorithm in* CRAY-FORTRAN, ACM Trans. Math. Software, 10, pp. 221–230.

J. DONGARRA, F. GUSTAVSON, AND A. KARP, [1984], *Implementing linear algebra algorithms for dense matrices on a vector pipeline machine*, SIAM Rev., 26, pp. 91–112.

J. DONGARRA AND A. HINDS, [1979], *Unrolling loops in* FORTRAN, Softw. Pract. Exper., 9, pp. 219–229.

J. DONGARRA AND R. HIROMOTO, [1984], *A collection of parallel linear equation routines for the Denelcor* HEP, Parallel Computing, 1, pp. 133–142.

J. DONGARRA AND A. SAMEH, [1984], *On some parallel banded system solvers*, Argonne National Laboratory Report ANL/MCS-TM-27, Argonne, IL.

J. DONGARRA, A. SAMEH AND D. SORENSON, [1984], *Implementation of some concurrent algorithms for matrix factorization*, Argonne National Laboratory Report. ANL/MCS-TM-25, Argonne, IL.

F. DORR, [1970], *The direct solution of the discrete Poisson equation on a rectangle*, SIAM Rev., 12, pp. 248–263.

R. DRESSLER, S. ROBERTSON, AND L. SPRADLEY, [1982], *Effects of Rayleigh accelerations applied to an initially moving fluid*, Materials Processing in the Reduced Gravity Environment of Space, G. Rindonne, ed., Elsevier, New York.

J. DRUMMOND, [1983], *Numerical study of a ramjet dump combustor flow field*, AIAA Paper No. 83-0421.

J. DRUMMOND AND E. WEIDNER, [1982], *Numerical study of a scramjet engine flow field*, AIAA J., 20, pp. 1182–1187.

M. DUBOIS AND F. BRIGGS, [1982], *Performance of synchronized iterative processes in multiprocessor systems*, IEEE Trans. Softw. Eng. Se-8, pp. 419–431.

P. DUBOIS, [1982], *Swimming upstream: table lookups and the evaluation of piecewise defined functions on vector computers*, in Rodrigue [1982], pp. 129–151.

P. DUBOIS, A. GREENBAUM AND G. RODRIGUE, [1979], *Approximating the inverse of a matrix for use in iterative algorithms on vector processors*, Computing, 22, pp. 257–268.

P. DUBOIS AND G. RODRIGUE, [1977a], *An analysis of the recursive doubling algorithm*, in Kuck, et al., [1977], pp. 299–305.

P. DUBOIS AND G. RODRIGUE, [1977b], *Operator splitting on the* STAR *without transposing*, Lawrence Livermore National Laboratory Report No. UCID-17515, Livermore, CA.

I. DUFF, [1977], MA28—*a set of Fortran subroutines for sparse unsymmetric linear equations*, AERE Report No. R8730, Harwell, England.

I. DUFF, [1982a], *The solution of sparse linear equations on the* CRAY-1, CRAY Channels, 4, 3, pp. 4–9.

I. DUFF, [1982b], *The solution of sparse linear equations on the* CRAY-1, in CRAY Research, Inc. [1982], pp. 17–39.

I. DUFF, [1984], *The solution of sparse linear equations on the* CRAY-1, in Kowalik [1984], pp. 293–309.

I. DUFF AND J. REID, [1982], *Experience of sparse matrix codes on the* CRAY-1, Comp. Phys. Comm., 76, pp. 293–302.

R. DUGAN, I. DURHAM AND S. TALUKDAR, [1979], *An algorithm for power system simulation by parallel processing*, Proc. IEEE Power Eng. Soc. Summer Meeting.

M. DUNGWORTH, [1979], *The* CRAY-1 *computer system*, in Jesshope and Hockney [1979], vol. 2, pp. 51–76.

I. DURHAM, R. DUGAN, A. JONES AND S. TALUKDAR, [1979], *Power system simulation on a multiprocessor*, Proc. IEEE Power Eng. Soc. Summer Meeting.

J. EASTWOOD AND C. JESSHOPE, [1977], *The solution of elliptic partial differential equations using number theoretical transforms with applications to narrow or computer hardware*, Comp. Phys. Comm., 13, pp. 233–239.

D. EBERHARDT, D. BAGANOFF AND K. STEVENS, [1984], *Study of the mapping of Navier–Stokes algorithms onto multiple-instruction/multiple-data-steam computers*, NASA TM-85945, NASA Ames Research Center, Moffett Field, CA.

J. ECKERT, JR., J. MAUCHLY, H. GOLDSTEIN AND J. BRAINERD, [1945], *Description of the* ENIAC *and comments on electronic digital computing machines*, Applied Mathematics Panel Report No. 171. 2R, Univ. Pennsylvania, Philadelphia.

S. EISENSTAT AND M. SCHULTZ, [1981], *Trends in elliptic problem solvers*, in Schultz [1981], pp. 99–114.

H. ENGELI, T. GINSBURG, H. RUTHISHAUSER AND E. STIEFEL, [1959], *Refined iterative methods for computation of the solution and the eigenvalues of self-adjoint boundary value problems*, Mitteilungen aus dem Institut fur Angewandte Mathematik, 8, Birkhauser Verlag, Basel, Stuttgart.

M. ENSELME, C. FRABOUL, AND P. LECA, [1984], *An* MIMD *architecture system for* PDE *numerical simulation*, in Vichnevetsky and Stepleman [1984], pp. 502–509.

P. ENSLOW, [1977], *Multiprocessor organization: A survey*, Comp. Surveys, 9, pp. 103–129.

J. ERHEL, [1983], *Parallelisation d'un algorithme de gradient conjugué préconditionné*, INRIA Report 189.

J. Erhel, W. Jalby, A. Lichnewsky and F. Thomasset, [1983], *Quelques progres en calcul parallele et vectoriel*, Coll. Inf. Ser des Méthodes de Calcul Scientifique et Technique.

J. Erhel, A. Lichnewsky and F.Thomasset,[1982], *Parallelism in finite element computations*, presented at the IBM Symposium on Vector Computers and Scientific Computing, Rome, 1982.

J. Ericksen, [1972], *Iterative and direct methods for solving Poisson's equation and their adaptability to ILLIAC IV*, Center for Advanced Computation Document 60, Univ Illinois, Urbana–Champaign .

J. Ericksen and R. Wilhelmson, [1976], *Implementation of a convective problem requiring auxiliary storage*, ACM Trans. Math. Software, 2, pp. 187–195.

C. Ethridge, J. Moore and V. Trujillo, [1983], *Experimental parallel microprocessor system*, Los Alamos National Laboratory Report No. LA-UR-83-1676, Los Alamos, NM.

D. Evans, [1979], *On the numerical solution of sparse systems of finite element equations*, in The Mathematics of Finite Elements & Applications III, Mafelap 1978 Conference Proceedings, J. R. Whiteman, ed., Academic Press, New York, pp. 448–458.

D. Evans, ed., [1982a], *Parallel Processing Systems*, Cambridge Univ. Press, Cambridge.

D. Evans, [1982b], *Parallel numerical algorithms for linear systems*, in Evans [1982a], pp. 357–384.

D. Evans, [1983], *New parallel algorithms in linear algebra*, EDF—Bulletin de la Direction des Études et des Recherches—Ser C, 1, pp. 61–69.

D. Evans, [1984], *Parallel SOR iterative methods*, Parallel Computing, 1, pp. 3–18.

D. Evans and R. Dunbar, [1983], *The parallel solution of triangular systems of equations*, IEEE Trans. Comput., C-32, pp. 201–204.

D. Evans and A. Hadjidimos, [1980], *A modification of the quadrant interlocking factorisation parallel method*, Int. J. Comput. Math., 8, pp. 149–166.

D. Evans, A. Hadjidimos and D. Noutsos, [1981], *The parallel solution of banded linear equations by the new quadrant interlocking factorisation (QIF) method*, Int. J. Comput. Math., 9, pp. 151–162.

D. Evans and M. Hatzopoulos, [1979], *A parallel linear systems solver*, Int. J. Comput. Math., 7, pp. 227–238.

D. Evans and S. Okolie, [1981], *A recursive decoupling algorithm for solving banded linear systems*, Int. J. Comput. Math., 10, pp. 139–152.

D. Evans and R. Sojoodi-Haghighi, [1982], *Parallel iterative methods for solving linear equations*, Int. J. Comput. Math., 11, pp. 247–284.

D. Evans and N. Yousif, *Asynchronous and synchronous iterative methods for solving linear equations*, to appear.

V. Faber, [1981], *Block relaxation strategies*, in Schultz [1981], pp. 271–275.

E. Fadden, [1980], *The AD-10: a digital computer approach to time critical simulation*, Proc. 4th Power Plant Dynamics, Control, and Testing Symposium.

V. Faddeeva and D. Faddeev, [1977], *Parallel computations in linear algebra*, Kibernetica, 6, pp. 28–40.

P. Farmwald, [1984], *The S-1 Mark IIA supercomputer*, in Kowalik [1984], pp. 145–155.

G. Feierbach and D. Stevenson, [1979], *The ILLIAC IV*, in Jesshope and Hockney [1979], vol. 2, pp. 77–92.

M. Feilmeier, ed., [1977], *Parallel computers—parallel mathematics*, Int. Assoc. for Mathematics and Computers in Simulation.

M. Feilmeier, [1982], *Parallel numerical algorithms*, in Evans [1982a], pp. 285–338.

M. Feilmeier, G. Joubert and U. Schendel, eds., [1984], *Parallel Computing 83: Proceedings of the International Conference on Parallel Computing*, North-Holland, New York.

C. Felippa, [1981], *Architecture of a distributed analysis network for computational mechanics*, Computers and Structures, 13, pp. 405–413.

T. Feng, [1981], *A survey of interconnection networks*, Computer, 14, 12, pp. 12–27.

W. Fichtner, L. Nagel, R. Penumalli, W. Peterson and J. D'Arcy, [1984], *The impact of supercomputers on IC technology development and design*, Proc. IEEE, 72, pp. 76–112.

J. Field, A. Kapauan and L. Snyder, [1983], *Pringle: A parallel processor to emulate CHiP computers*, Computer Science Dept. Report No. CSD-TR-433, Purdue University, W. Lafayette, IN.

P. Flanders, D. Hunt, S. Reddaway and D. Parkinson, [1977], *Efficient high speed computing with the distributed array processor*, in Kuck, et al. [1977], pp. 113–128.

Floating Point Systems, [1976], *AP-120B Array Processor Handbook*, Floating Point Systems Reference Manual 7259-02.

M. Flynn, [1966]. *Very high speed computing systems*, Proc. IEEE, 54, pp. 1901–1909.

M. Flynn, [1972]. *Some computer organizations and their effectiveness*, IEEE Trans. Comput., C-21, pp. 948–960.

H. Foerster, K. Steuben and U. Trottenberg, [1981]. *Nonstandard multigrid techniques using checkered relaxation and intermediate grids*, in Schultz [1981], pp. 285–300.

K. Fong and T. Jordan, [1977], *Some linear algebraic algorithms and their performance on the* CRAY-1. Los Alamos National Laboratory Report LA-6774, Los Alamos, NM.

B. Fornberg, [1981], *A vector implementation of the fast Fourier transform algorithm*, Math. Comp., 36, pp. 189–191.

B. Fornberg, [1983], *Steady viscous flow past a circular cylinder*, in Gary [1984], pp. 201–224.

C. Foster, [1976], *Content Addressable Parallel Processors*, Van Nostrand–Reinhold, New York.

G. Fox, [1984], *Concurrent processing for scientific calculations*, Proc. COMPCON 84, IEEE Comp. Sci. Conf., pp. 70–73.

G. Fox and S. Otto, [1984], *Algorithms for concurrent processors*, Physics Today, 37, 5, pp. 50–59.

M. Franklin, [1978], *Parallel solution of ordinary differential equations*, IEEE Trans. Comput. C-25, pp. 413–470.

A. Friedman and D. Kershaw, [1982], *Vectorized incomplete Cholesky conjugate gradient* (ICCG) *package for the* CRAY-1 *computer*, Laser Program Annual Report UCRL-500021-81, Lawrence Livermore Nat. Lab., Livermore, CA.

S. Fuller, A. Jones and I. Durham, [1980], CMU Cm* *review*, Computer Science Dept. Report AD-A050135, Carnegie–Mellon University, Pittsburgh.

S. Fuller and P. Oleinick, [1976], *Initial measurements of parallel programs on a multi-miniprocessor*, Proc. 13th IEEE Computer Soc. Int. Conf., pp. 358–363.

S. Fuller, J. Ousterbout, L. Raskin, P. Rubinfeld, P. Sundhu, and R. Swan, [1978], *Multi-microprocessors: an overview and working example*, Proc. IEEE, 66, pp. 216–228.

D. Gajski, [1979], *Solving banded triangular systems on pipelined machines*, Proc. 1979 Int. Conf. Par. Proc., pp. 308–319.

D. Gajski, [1981], *An algorithm for solving linear recurrence systems on parallel and pipelined machines*, IEEE Trans. Comput., C-30, pp. 190–206.

D. Gajski, D. Kuck, D. Lawrie and A. Sameh, [1983], *Cedar—a large scale multiprocessor*, Proc. 1983 Int. Conf. Par. Proc., pp. 524–529.

D. Gajski, D. Lawrie, D. Kuck and A. Sameh, [1984], *Cedar*, Proc. COMPCON 84, IEEE Comp. Soc. Conf., pp. 306–309.

D. Gajski, A. Sameh and J. Wisniewski, [1982], *Iterative algorithms for tridiagonal matrices on a* WSI-*multiprocessor*, Proc. 1982 Int. Conf. Par. Proc., pp. 82–89.

Z. Galil and W. Pauli, [1983], *An efficient general-purpose parallel computer*, J. ACM, 30, pp. 286–299.

E. Gallopoulos, [1984], *The massively parallel processor for problems in fluid dynamics*, Proc. Vector and Parallel Processors in Computational Science II Conference, Oxford, England.

E. Gallopoulos and S. McEwan, [1983], *Numerical experiments with the massively parallel processor*, Proc. 1983 Int. Conf. Par. Proc., pp. 29–35.

D. Gannon, [1980], *A note on pipelining a mesh-connected multiprocessor for finite element problems by nested dissection*, Proc. 1980 Int. Conf. Par. Proc., pp. 197–204.

D. Gannon, [1981], *On mapping non-uniform PDE structures and algorithms onto uniform array architectures*, Proc. 1981 Int. Conf. Par. Proc., pp. 100–105.

D. Gannon and J. Panetta, [1985], *Restructuring SIMPLE for the CHIP architecture*, Parallel Computing, 2, to appear.

D. Gannon, L. Snyder and J. Van Rosendale, [1983], *Programming substructure computations for elliptic problems on the CHIP system*, in Noor [1983], pp. 65–80.

D. Gannon and J. Van Rosendale, [1982], *Highly parallel multigrid solvers for elliptic PDE's: an experimental analysis*, ICASE Report No. 82-36, NASA Langley Research Center, Hampton, VA.

D. Gannon and J. Van Rosendale, [1984a], *Parallel architectures for iterative methods on adaptive, block structured grids*, in Birkhoff and Schoenstadt [1984], pp. 93–104.

D. Gannon and J. Van Rosendale, [1984b], *On the impact of communication complexity in the design of parallel numerical algorithms*, IEEE Trans. Comput., C-33, pp. 1180–1194.

Q-S. Gao and R.-Q. Wang, [1983], *Vector computer for sparse matrix operations*, Proc 1983 Int. Conf. Par. Proc., pp. 87–89.

J. Gary, ed., [1984], CYBER 200 *applications seminar*, Proc. Seminar held at NASA Goddard Space Flight Center, October, 1983, NASA-CP-2295.

J. Gary, [1977], *Analysis of applications programs and software requirements for high speed computers*, in Kuck, et al. [1977], pp. 329–354.

J. Gary, S. McCormick and R. Sweet, [1983], *Successive overrelaxation, multigrid, and preconditioned conjugate gradient algorithms for solving a diffusion problem on a vector computer*, Appl. Math. Comp. 13, pp. 285–309.

M. Gautzsch, G. Weiland and D. Muller-Richards, [1980], *Possibilities and problems with the application of vector computers*, German Research and Testing Establishment for Aerospace.

E. GEHRINGER, A. JONES AND Z. SEGALL, [1982], *The Cm* testbed*, Computer, 15, 10, p. 40–53.

E. GELENBE, A. LICHNEWSKY AND A. STAPHYLOPATIS, [1982], *Experience with the parallel solution of partial differential equations on a distributed computing system*, IEEE Trans. Comput., C-31, pp. 1157–1165.

W. GENTLEMAN, [1975], *Error analysis of the QR decomposition by Givens transformations*, Lin. Alg. Appl., 10, pp. 189–197.

W. GENTLEMAN, [1978], *Some complexity results for matrix computations on parallel processors*. J. ACM, 25, pp. 112–115.

W. GENTLEMAN, [1981], *Design of numerical algorithms for parallel processing*, presented at the Parallel Processing Conference at Bergams, Italy.

W. GENTLEMAN AND H. KUNG, [1981], *Matrix triangularization by systolic arrays*, Proc. SPIE 298, Real-Time Signal Processing IV, pp. 19–26.

W. GENTZSCH, [1983], *How to maintain the efficiency of highly serial algorithms involving recursions on vector computers*, Proc. Conf. Vector and Parallel Methods in Scientific Computing, Paris.

W. GENTZSCH, [1984a], *Benchmark results on physical flow problems*, in Kowalik [1984], pp. 211–228.

W. GENTZSCH, [1984b], *Vectorization of Computer Programs With Applications to Computational Fluid Dynamics*, Heyden & Son, Philadelphia.

W. GENTZSCH, [1984c], *Numerical algorithms in computational fluid dynamics on vector computers*, Parallel Computing, 1, pp. 19–33.

A. GEORGE, [1972], *An efficient band-oriented scheme for solving n by n grid problems*, Proc. 1972 FJCC, AFIPS Press, Montvale, NJ, pp. 1317–1321.

A. GEORGE, [1973], *Nested dissection of a regular finite element mesh*, SIAM J. Numer. Anal., 10, pp. 345–363.

A. GEORGE, [1977], *Numerical experiments using dissection methods to solve n by n grid problems*, SIAM J. Numer. Anal., 14, pp. 161–179.

A. GEORGE AND J. LIU, [1981], *Computer Solution of Large Sparse Positive Definite Systems*, Prentice-Hall, Englewood Cliffs, NJ.

A. GEORGE, W. POOLE AND R. VOIGT, [1978a], *A variant of nested dissection for solving n by n grid problems*, SIAM J. Numer. Anal., 15, pp. 662–673.

A. GEORGE, W. POOLE AND R. VOIGT, [1978b], *Analysis of dissection algorithms for vector computers*, Comput. Math. Appl., 4, pp. 287–304.

P. GILMORE, [1971], *Numerical solution of partial differential equations by associative processing*, Proc. 1971 FJCC, AFIPS Press, Montvale, NJ, pp. 411–418.

M. GINSBURG, [1982], *Some observations on supercomputer computational environments*, Proc. 10th IMACS World Congress on Systems Simulation and Scientific Computation, vol. 1, IMACS, New Brunswick, NJ, pp. 297–301.

E. GIROUX, [1977], *A large mathematical model implementation on the STAR-100 computer*, in Kuck, et al. [1977], pp. 287–298.

I. GLOUDEMAN, [1984], *The anticipated impact of supercomputers on finite element analysis*, Proc IEEE, 72, pp. 80–84.

J. GLOUDEMAN, C. HENNRICH AND J. HODGE, [1984], *The evolution of MSC/NASTRAN and the supercomputer for enhanced performance*, in Kowalik [1984], pp. 393–402.

J. GLOUDEMAN AND J. HODGE, [1982], *The adaption of MSC/NASTRAN to a supercomputer*, Proc. 10th IMACS World Congress on Systems Simulation and Scientific Computation, vol. 1, IMACS, New Brunswick, NJ, pp. 302–304.

P. GNOFFO, [1982], *A vectorized, finite-volume, adaptive-grid algorithm for Navier–Stokes calculations*, in Numerical Grid Generation, J. Thompson, ed., Elsevier, New York.

R. GOKE AND G. LIPOVSKI, [1973], *Banyan networks for partitioning on multiprocessor systems*, Proc. 1st Ann. Symp. Computer Arch., pp. 21–30.

G. GOLUB AND D. MAYERS, [1983], *The use of preconditioning over irregular regions*, Numerical Analysis Project Report No. NA-83-27, Stanford Univ., Stanford, CA.

Goodyear Aerospace Corp. [1974], *Application of STARAN to fast Fourier transforms*, Report GER 16109, May.

K. GOSTELOW AND R. THOMAS, [1980], *Performance of a simulated dataflow computer*, IEEE Trans. Comput. C-29, pp. 905–919.

A. GOTTLIEB, [1984], *Avoiding serial bottlenecks in ultraparallel MIMD computers*, Proc. COMPCON 84, IEEE Comp. Soc. Conf., pp. 354–359.

A. GOTTLIEB, R. GRISHMAN, C. KRUSKAL, K. McAULIFFE, L. RUDOLPH AND M. SNIR, [1983], *The NYU ultracomputer—designing an MIMD shared memory-parallel computer*, IEEE Trans. Comput., C-32, pp. 175–189.

A. GOTTLIEB, B. LUBACHEVSKY AND L. RUDOLPH, [1983], *Basic techniques for the efficient coordination of very large numbers of cooperating sequential processors*, ACM Trans. Program. Lang. Syst., 5, pp. 164–189.

A. GOTTLIEB AND J. SCHWARTZ, [1982], *Networks and algorithms for very-large-scale parallel computation*, Computer, 15, No. 1, pp. 27–36.

D. GOTTLIEB, M. HUSSAINI AND S. ORSZAG, [1984], *Theory and applications of spectral methods*, in Voigt, et al. [1984], pp. 1–54.

D. GOTTLIEB AND S. ORSZAG, [1977], *Numerical Analysis of Spectral Methods: Theory and Applications*, CBMS Regional Conference Series in Applied Mathematics 26, Society for Industrial and Applied Mathematics, Philadelphia.

D. GOTTLIEB AND E. TURKEL, [1976], *Boundary conditions for multistep finite difference methods for time dependent equations*, J. Comp. Phys., 26, pp. 181–196.

G. L. GOUDREAU, R. A. BAILEY, J. O. HALLQUIST, R. C. MURRAY AND S. J. SACKETT, [1983], *Efficient large scale finite element computations in a Cray environment*, in Noor [1983], pp. 141–154.

M. GRAHAM, [1976], *An array computer for the class of problems typified by the general circulation model of the atmosphere*, Ph. D. thesis, Dept. Computer Science, University of Illinois, Urbana–Champaign.

R. GRAVES, [1973], *Partial implicitization*, J. Comp. Phys., 13, pp. 439–444.

J. GRCAR AND A. SAMEH, [1981], *On certain parallel Toeplitz linear system solvers*, SIAM J. Sci. Stat. Comput., 2, pp. 238–256.

A. GREENBAUM AND G. RODRIGUE, [1977], *The incomplete Choleski conjugate gradient method for the STAR (5 point operator)*, Lawrence Livermore National Laboratory Report, Livermore, CA.

D. GRIT AND J. McGRAW, [1983], *Programming divide and conquer on a multiprocessor*, Lawrence Livermore National Laboratory Report UCRL-88710, Livermore, CA.

C. GROSCH, [1978], *Poisson solvers on a large array computer*, Proc. 1978 LASL Workshop on Vector and Parallel Processors, pp. 98–132.

C. GROSCH, [1979a]. *Performance analysis of tridiagonal equation solvers on array computers*, Dept. Mathematical and Computing Sciences Technical Report No. TR 79-4, Old Dominion University, Norfolk, VA.

C. GROSCH, [1979b], *Performance analysis of Poisson solvers on array computers*, in Jesshope and Hockney [1979], vol. 2, pp. 147–181.

C. GROSCH, [1980], *The effect of the data transfer pattern of an array computer on the efficiency of some algorithms for the tridiagonal and Poisson problems*, presented at the Conference on Array Architectures for Computing in the 80's and 90's, Hampton, VA.

R. GUILILAND, [1981], *Solution of the shallow water equations on the sphere*, J. Comp. Phys., 43, pp. 79–94.

J. GURD AND I. WATSON, [1982], *Preliminary evaluation of a prototype dataflow computer*, Proc. IFIP World Computer Congress, North-Holland, Amsterdam, pp. 545–551.

W. HACKBUSCH, [1978], *On the multigrid method applied to difference equations*, Computing, 20, pp. 291–306.

W. HACKBUSCH AND U. TROTTENBERG, eds., [1982], *Multigrid Methods*, Springer-Verlag, Berlin.

M. HAFEZ AND D. LOVELL, [1983], *Improved relaxation schemes for transonic potential calculations*, AIAA Paper 83-0372.

M. HAFEZ AND E. MURMAN, [1978], *Artificial compressibility methods for numerical solution of transonic full potential equations*, AIAA 11th Fluid and Plasma Dynamics Conference, Seattle, WA.

M. HAFEZ AND J. SOUTH, [1979], *Vectorization of relaxation methods for solving transonic full potential equations*, Flow Research Report, Flow Research, Inc., Kent, WA.

L. HALADA, [1980], *A parallel algorithm for solving band systems of linear equations*, Proc. 1980 Int. Conf. Par. Proc., pp. 159–160.

L. HALADA, [1981], *A parallel algorithm for solving band systems and matrix inversion*, CONPAR 81, Conf. Proc., Lecture Notes in Computer Science, W. Handler, ed., Springer-Verlag, pp. 433–440.

H. HALIN, R. BUHRER, W. HALG, H. BENZ, B. BRON, H. BRUNDIERS, A. ISAACSON, AND M. TADIAN, [1980], *The ETHM multiprocessor project: parallel simulation of continuous systems*, Simulation, 35, pp. 109–123.

W. HANDLER, E. HOFMANN AND H. SCHNEIDER, [1976], *A general purpose array with a broad spectrum of applications*, in Informatik-Fachberichte, Springer-Verlag, Berlin.

W. HANKEY AND J. SHANG, [1982], *Vector processors and CFD*, in Cray Research, Inc. [1982], pp. 49–66.

H. HAPP, C. POTTE AND K. WIRGAN, [1978], *Parallel processing for large scale transient stability*, Proc. IEEE Can. Conf. Comm. Power, pp. 204–207.

A. HARDING AND J. CARLING, [1984], *The three-dimensional solution of the equations of flow and heat transfer in glass-melting tank furnaces: adapting to the DAP*, in Paddon [1984], pp. 115–133.

M. HATZOPOULOS, [1982], *Parallel linear system solvers for tridiagonal matrices*, in Evans [1982a], pp. 389–394.

L. HAYES, [1974], *Comparative analysis of iterative techiques for solving Laplace's equation on the unit square on a parallel processor*, M. S. Thesis, Dept., Mathematics, Univ. Texas, Austin.

L. HAYES AND P. DEVLOO, [1984], *An overlapping block iterative scheme for finite element methods*, Dept. Aerospace Engineering and Engineering Mechanics, Univ. Texas, Austin.

L. HAYNES, R. LAU, D. SIEWIOREK, AND D. MIZELL, [1982], *A survey of highly parallel computing*, Computer, 15, 1, pp. 9–24.

D. HELLER, [1974], *A determinant theorem with applications to parallel algorithms*, SIAM J. Numer. Anal., 11, pp. 559–568.

D. HELLER, [1976], *Some aspects of the cyclic reduction algorithm for block tridiagonal linear systems*, SIAM J. Numer. Anal., 13, pp. 484–496.

D. HELLER, [1978], *A survey of parallel algorithms in numerical linear algebra*, SIAM Rev., 20, pp. 740–777.

D. HELLER AND I. IPSEN, [1983], *Systolic networks for orthogonal decompositions*, SIAM J. Sci. Stat. Comput., 4, pp. 261–269.

D. HELLER, D. STEVENSON AND J. TRAUB, [1976], *Accelerated iterative methods for the solution of tridiagonal linear systems on parallel computers*, J. ACM, 23, pp. 636–654.

R. HELLIER, [1982], DAP *implementation of the WZ algorithm*, Comp. Phys., Comm. 321–323.

P. HEMKER, R. KETTLER, P. WESSELING, AND P. DE ZEEUW, [1983], *Multigrid methods: development of fast solvers*, Appl. Math. Comp., 13, pp. 311–326.

J. HENDRY AND L. DELVES, [1984], GEM *calculations on the DAP*, in Paddon [1984], pp. 185–194.

L. HERTZBERGER, D. GOSMAAN, G. KIEFT, M. SCHOOREL, AND L. WIGGERS, [1981], FAMP *system*, Comp. Phys. Comm., 22, pp. 253–260.

M. HESTENES AND E. STIEFEL, [1952], *Methods of conjugate gradients for solving linear systems*, J. Res. Nat. Bur. Standards Sect. B, 49, pp. 409–436.

P. HIBBARD AND N. OSTLUND, [1980], *Numerical computation on* Cm*, Proc. 1980 Int. Conf. Par. Proc., pp. 135–136.

L. HIGBIE, [1978], *Speeding up* FORTRAN (CFT) *programs on the* CRAY-1, CRAY Research Inc. Pub. 2240207.

R. HINTZ AND D. TOTE, [1972], *Control Data* STAR-100 *processor design*, Proc. COMPCON 72, IEEE Comp. Soc. Conf., pp. 1–4.

K. HIRAKI, T. SHIMADA, AND K. NISHIDA, [1984], *A hardware design of the* SIGMA-1, *a data flow computer for scientific computations*, Proc. 1984 Int. Conf. Par. Proc., pp. 524–531.

L. HOBBS, D. THEIS, J. TRIMBLE, H. TITUS, AND D. HIGHBERG, [1970], *Parallel Processor Systems: Technologies and Applications*, Spartan Books, New York.

R. HOCKNEY, [1965], *A fast direct solution of Poisson's equation using Fourier analysis*, J. ACM, 12, pp. 95–113.

R. HOCKNEY, [1970], *The potential calculation and some applications*, Meth. Comput. Phys., 9, pp. 135–211.

R. HOCKNEY, [1977], *Super-computer architecture*, Proc. Infotech State of the Art Conf. on Future Systems.

R. HOCKNEY, [1979], *The large parallel computer and university research*, Cont. Phys., 20, pp. 149–185.

R. HOCKNEY, [1982a], *Optimizing the* FACR(l) *Poisson solver on parallel computers*, Proc. 1982 Int. Conf. Par. Proc., pp. 62–71.

R. HOCKNEY, [1982b], *Poisson solving on parallel computers*, presented at the IBM Symposium on Vector Computers and Scientific Computing, Rome.

R. HOCKNEY, [1982c], *Characterization of parallel computers and algorithms*, Comp. Phys. Comm., 26, pp. 285–291.

R. HOCKNEY, [1983a], *Characterization of parallel computers*, Proc. of World Congress on System Simulation and Scientific Computation, International Association for Mathematics and Computers in Simulation, vol. 1, pp. 269–271.

R. HOCKNEY, [1983b], *Characterizing computers and optimizing the* FACR(l) *Poisson solver on parallel unicomputers*, IEEE Trans. Comp., C32, pp. 933–941.

R. HOCKNEY, [1984a], *Performance of parallel computers*, in Kowalik [1984], pp. 159–176.

R. HOCKNEY, [1984b], *Optimizing the* FACR(l) *Poisson-solver on parallel computers*, in Paddon [1984], pp. 45–65.

R. HOCKNEY AND C. JESSHOPE, [1981], *Parallel Computers: Architecture, Programming and Algorithms*, Adam Hilger, Bristol.

J. HOLLAND, [1959], *A universal computer capable of executing an arbitrary number of subprograms simultaneously*, Proc. European Joint Comp. Conf., pp. 108–113.

R. HORD, [1982], *The Illiac* IV: *The First Supercomputer*, Computer Science Press, Potomac, MD.

T. HOSHINO, T. KAWAI, T. SHIRAKAWA, J. HIGASHINO, A. YAMAOKA, H. ITO, T. SATO, AND K. SAWADA, [1983], PACS: *A parallel microprocessor array for scientific calculations*, ACM Trans. Comp. Sys. 1, pp. 195–221.

T. Hoshino, T. Shirakawa, T. Kamimura, T. Kageyama, K. Takenouochi, H. Abe, S. Sekiguchi, Y. Oyanagi, and K. Toshio, [1983], *Highly parallel processor array "PAX" for wide scientific applications*, Proc. 1983 Int. Conf. Par. Proc., pp. 95–105.

S. Hotovy and L. Dickson, [1979], *Evaluation of a vectorizable 2-D transonic finite difference algorithm*, AIAA Paper 79-0276.

E. Housors, and O. Wing, [1984], *Pseudo-conjugate directions for the solution of the nonlinear unconstrained optimization problem on a parallel computer*, J. Optim. Theory Appl., 42, pp. 169–180.

J. Huang and O. Wing, [1979], *Optimal parallel triangulation of a sparse matrix*, IEEE Trans. Circuits Syst., CAS-26, pp. 726–732.

K. Huang and J. Abraham, [1982], *Efficient parallel algorithms for processor arrays*, Proc.1982 Int. Conf. Par. Proc., pp. 271–279.

K. Huang and J. Abraham, [1984], *Fault-tolerant algorithms and their application to solving Laplace equations*, Proc. 1984 Int. Conf. Par. Proc., pp. 117–122.

R. Huff, J. Dawson and G. Culler, [1982], *Plasma physics on an array processor*, in Rodrigue [1982], pp. 365–396.

D. Hunt, [1979], *Application techniques for parallel hardware*, in Jesshope and Hockney [1979], pp. 205–219.

D. Hunt, S. Webb, and A. Wilson, [1981], *Applications of a parallel processor to the solution of finite difference problems*, in Schultz [1981], pp. 339–344.

K. Hwang, [1982], *Partitioned matrix algorithms for VLSI arithmetic systems*, IEEE Trans. Comput., C-31, pp. 1215–1224.

K. Hwang and F. Briggs, [1984], *Computer Architecture and Parallel Processing*, McGraw-Hill, New York.

K. Hwang and Y.-H. Cheng, [1980], *VLSI computing structures for solving large scale linear systems of equations*, Proc. 1980 Int. Conf. Par. Proc., pp. 217–227.

K. Hwang, S. Su, and L. Ni, [1981], *Vector computer architecture and processing techniques*, Advances in Computers 20, pp. 115–197.

L. Hyafil and H. Kung, [1974], *Parallel algorithms for solving triangular linear systems with small parallelism*, Dept. Computer Science Report, Carnegie-Mellon Univ., Pittsburgh.

L. Hyafil and H. Kung, [1975], *Bounds on the speed-ups of parallel evaluation of recurrences*, Proc. Second USA-Japan Comp. Conf., pp. 178–182.

L. Hyafil and H. Kung, [1977], *The complexity of parallel evaluation of linear recurrences*, J. ACM, 24, pp. 513–521.

IBM Corp., [1978], *Vector processing subsystem (VPSS) programmers guide*, IBM Ref. Manual GC24-3716-0.

M. Inouye, ed., [1977], *Future computer requirements for computational aerodynamics*, Workshop at NASA-Ames Research Center, Conf. Publ. No. 2032.

I. Ipsen, [1984], *A parallel QR method using fast Givens rotations*, Dept. Computer Science Report RR 299, Yale Univ., New Haven, CT.

I. Ipsen, Y. Saad, and M. Schultz, [1985], *Complexity of dense linear system solution on a multiprocessor ring*, Dept. Computer Science Report RR-349, Yale Univ., New Haven, CT.

A. Jameson and E. Turkel, [1979], *Implicit schemes and LU decompositions*, Math. Comp., 37, pp. 385–397.

A. Jennings, [1966], *A compact storage scheme for the solution of symmetric linear simultaneous equations*, Computer J., 9, pp. 281–285.

J. Jess and H. Kees, [1982], *A data structure for parallel L/U decomposition*, IEEE Trans. Comput., C-31, pp. 231–239.

C. Jesshope, [1977], *Evaluation of Illiac: overlap, nonoverlap*, Institute for Advanced Computation Newsletter, vol. 1, pp. 4–5.

C. Jesshope, [1980a], *The implementation of the fast radix 2 transforms on array processors*, IEEE Trans. Comput., C-29, pp. 20–27.

C. Jesshope, [1980b], *Some results concerning data routing in array processors*, IEEE Trans. Comput., C-29, pp. 659–662.

C. Jesshope and J. Craigie, [1979], *Some principles of parallelism in particle and mesh modelling*, in Jesshope and Hockney [1979], vol. 2, pp. 221–236.

C. Jesshope and R. Hockney, eds., [1979], *Infotech State of the Art Report: Supercomputers*, Vols. 1 & 2, Infotech, Maidenhead, England.

J. Johnson, [1983], *ETA leaves home*, Datamation, 29, 10, pp. 74–86.

O. Johnson, [1981], *Vector function chainer software for banded preconditioned conjugate gradient calculations*, Advances in Computer Methods for Partial Differential Equations - IX, Proc. 10th IMACS World Congress on Systems Simulation and Scientific Computation, vol. 1, IMACS, New Brunswick, NJ, pp. 243–245.

O. Johnson, [1984], *Three-dimensional wave equation computations on vector computers*, Proc. IEEE, 72, pp. 90–95.

O. JOHNSON AND M. EDWARDS, [1981], *Progress on the* 3D *wave equation program for the* CDC Cyber 205, Seismic Acoustics Lab., Fourth year Semi-Annual Prog. Rep., vol. 7, pp. 11–15.

O. JOHNSON AND M. LEWITT, [1982], PPCG *software for the* CDC CYBER 205, in Control Data Corp. [1982].

O. JOHNSON, C. MICCHELLI AND G. PAUL, [1983], *Polynomial preconditioners for conjugate gradient calculations*, SIAM J. Numer. Anal., 20, pp. 362–376.

O. JOHNSON AND G. PAUL, [1981a], *Optimal parametrized incomplete inverse preconditioning for conjugate gradient calculations*, IBM Report RC 8644, Yorktown Heights, NY.

O. JOHNSON AND G. PAUL, [1981b], *Vector algorithms for elliptic partial differential equations based on the Jacobi method*, in Schultz [1981], pp. 345–351.

L. JOHNSSON, [1984a], *Highly concurrent algorithms for solving linear systems of equations*, in Birkhoff and Schoenstadt [1984], pp. 105–126.

L. JOHNSSON, [1984b], *Odd-even cyclic reduction on ensemble architectures and the solution of tridiagonal systems of equations*, Dept. Computer Science Report YALEU/CSD/RR-339, Yale Univ., New Haven, CT.

A. JONES, R. CHANSLER, I. DURHAM, P. FEILER, D. SCELZA, K. SCHWANS AND S. VEGDAHL, [1978], *Programming issues raised by a multi-microprocessor*, Proc. IEEE, 66, pp. 229–237.

A. JONES AND E. GEHRINGER, eds., [1980], *The* Cm* *multiprocessor project: a research review*, Computer Science Dept. Report CMU-CS-80-131, Carnegie-Mellon Univ., Pittsburgh.

A. JONES AND P. SCHWARTZ, [1980], *Experience using multiprocessor systems: a status report*, Computing Surveys, 12, pp. 121–165.

H. JORDAN, [1978a], *A special purpose architecture for finite element analysis*, Proc. 1978 Int. Conf. Par. Proc., pp. 263–266.

H. JORDAN, [1978b], *The finite element machine programmer's reference manual*, Dept. Computer Science Report CSDG 78–2, Univ. Colorado, Boulder.

H. JORDAN, [1981], *Parallelizing a sparse matrix package*, Computer Systems Design Group Report CSDG 81–1, Univ. Colorado, Boulder.

H. JORDAN, [1983], *Performance measurements on* HEP - *a pipelined* MIMD *computer*, Proc. 10th. Ann. Int. Symp. Comp. Arch.

H. JORDAN, [1984], *Experience with pipelined multiple instruction streams*, Proc. IEEE, 72, pp. 113–123.

H. JORDAN AND D. PODSIADLO, [1980], *A conjugate gradient program for the finite element machine*, Dept. Computer Science Report CSDG, Univ. Colorado, Boulder.

H. JORDAN AND P. SAWYER, [1979], *A multimicroprocessor system for finite element structural analysis*, in Trends in Computerized Structural Analysis and Synthesis, A. Noor and H. McComb, eds., Pergamon, New York, pp. 21–29.

H. JORDAN, M. SCALABRIN AND W. CALVERT, [1979], *A comparison of three types of multiprocessor algorithms*, Proc. 1979 Int. Conf. Par. Proc., pp. 231–238.

T. JORDAN, [1974], *A new parallel algorithm for diagonally dominant tridiagonal matrices*, Los Alamos National Laboratory Report, Los Alamos, NM.

T. JORDAN, [1979], *A performance evaluation of linear algebra software in parallel architectures*, in Performance Evaluation of Numerical Software, L. Fosdick. ed., North-Holland, Amsterdam, pp. 59–76.

T. JORDAN, [1982a], *A guide to parallel computation and some* CRAY-1 *experiences*, in Rodrigue [1982], pp. 1–50.

T. JORDAN, [1982b], CALMATH: *Some problems and applications*, in Cray Research, Inc. [1982], pp. 5–8.

T. JORDAN, [1984], *Conjugate gradient preconditioners for vector and parallel processors*, in Birkhoff and Schoenstadt [1984], pp. 127–139.

T. JORDAN AND K. FONG, [1977], *Some linear algebraic algorithms and their performance on the* CRAY-1, in Kuck, et al. [1977], pp. 313–316.

E. KALNAY AND L. TAKOCS, [1982], *A simple atmospheric model on the sphere with* 100% *parallelism*, NASA-Goddard Modeling and Simulation Facility Research Review, 1980–81, pp. 89–95.

E. KALNAY-RIVAS, A. BAYLISS AND J. STORCH, [1976], *Experiments with the fourth order* GISS *model of the global atmospheric*, Proc. Conf. on Simulation of Large-Scale Atmospheric Processes, Hamburg, Germany.

C. KAMATH AND A. SAMEH, [1984], *The preconditioned conjugate gradient algorithm on a multiprocessor*, in Vichnevetsky and Stepleman [1984], pp. 210–217.

Y. KANEDA AND M. KOHATA, [1982], *Highly parallel computing of linear equations on the matrix-broadcast memory connected array processor system*, Proc. 10th IMACS World Congress on Systems Simulation and Scientific Computation, vol. 1, IMACS, New Brunswick, NJ, pp. 320–322.

R. KANT AND T. KIMURA, [1978], *Decentralized parallel algorithms for matrix computations*, Proc. 5th Annual Symp. Comp. Arch., pp. 96–100.

A. Kapauan, K. Wang, D. Gannon, J. Cuny and L. Snyder, [1984], *The Pringle: An experimental system for parallel algorithm and software testing*, Proc. 1984 Int. Conf. Par. Proc., pp. 1–6.

R. Kapur and J. Browne, [1981], *Block tridiagonal linear systems on a reconfigurable array computer*, Proc. 1981 Int. Conf. Par. Proc., pp. 92–99.

R. Kapur and J. Browne, [1984], *Techniques for solving block tridiagonal systems on reconfigurable array computers*, SIAM J. Sci. Stat. Comp., 5, pp. 701–719.

A. Kasahara, [1984], *Recent mathematical and computational developments in numerical weather prediction*, in Parter [1984], pp. 85–126.

M. Kascic, [1978], *A direct Poisson solver on STAR*, Proc. 1978 LASL Workshop on Vector and Parallel Processors.

M. Kascic, [1979a], *Vector processing on the CYBER 200*, in Jesshope and Hockney [1979], pp. 237–270.

M. Kascic, [1979b], *Vector processing on the CYBER 200 and vector numerical linear algebra*, Proc. 3rd GAMM Conf. on Numeric Mathematics in Fluid Dynamics.

M. Kascic, [1983a], *Anatomy of a Poisson solver*, Proc. Parallel 83 Conf., Berlin.

M. Kascic, [1983b], *Syntactic and semantic vectorization: whence cometh intelligence in supercomputing?* Proc. 1983 Summer Computer Simulation Conf., Vancouver.

M. Kascic, [1984a], *A performance survey of the CYBER 205*, in Kowalik [1984], pp. 191–210.

M. Kascic, [1984b], *Vorton dynamics: A case study of developing a fluid dynamics model for a vector processor*, Parallel Computing, 1, pp. 35–44.

M. Kascic, [1984c], *Interplay between computer methods and partial differential equations: Iterative methods as exemplar*, in Vichnevetsky and Stepleman [1984], pp. 379–382.

H. Kashiwagi, [1984], *Japanese super speed computer project*, in Kowalik [1984], pp. 117–125.

J. Keller and A. Jameson, [1978], *Preliminary study of the use of the STAR 100 computer for transonic flow calculations*, AIAA paper 78–12.

R. Kendall, G. Morrell, D. Peaceman, W. Silliman, and J. Watts, [1983], *Development of a multiple application reservoir simulator for use on a vector computer*, SPE Paper 11483, SPE Middle East Oil Tech. Conf., Bahrain.

R. Kendall, J. Nolen and P. Stanat, [1984], *The impact of vector processors on petroleum reservoir simulation*, Proc. IEEE, 72, pp. 85–89.

M. Kenichi, [1981], *A vector oriented finite-difference scheme for calculating three-dimensional compressible laminar and turbulent boundary layers on practical wing configurations*, AIAA Paper 81–1020.

D. Kershaw, [1982], *Solution of single tridiagonal linear systems and vectorization of the ICCG algorithm on the CRAY-1*, in Rodrigue [1982], pp. 85–89.

J. Killough, [1979], *The use of vector processors in reservoir simulation*, Proc. SPE Symposium Reservoir Simulation, Denver.

T. Kimura, [1979], *Gauss–Jordan elimination by VLSI mesh-connected processors*, in Jesshope and Hockney [1979], Vol. 2, pp. 271–290.

D. Kincaid, G. Carey, T. Oppe, K. Sepehenoori and D. Young, [1984], *Combining finite element and iterative methods for solving partial differential equations on advanced computer architectures*, in Vichnevetsky and Stepleman [1984], pp. 375–378.

D. Kincaid and T. Oppe, [1983], *ITPACK on supercomputers*, in Numerical Methods, A. Dold and B. Eckman, eds., Springer-Verlag, New York, 1983, pp. 151–161.

D. Kincaid, T. Oppe and D. Young, [1982], *Adapting ITPACK routines for use on a vector computer*, in Control Data Corp. [1982].

D. Kincaid, T. Oppe and D. Young, [1984], *Vector computations for sparse linear systems*, Center for Numerical Analysis Report CNA 189, Univ. Texas, Austin.

D. Kincaid and D. Young, [1983], *Adapting iterative algorithms for solving large sparse linear systems for efficient use of the CDC CYBER 205*, in Gary [1984], pp. 147–160.

D. Knight, [1983], *A hybrid explicit-implicit numerical algorithm for the three-dimensional compressible Navier-Stokes equations*, AIAA 21st Aerospace Sciences Meeting, January, Reno, NV. AIAA Paper No. 83–0223.

J. Knight and D. Dunlop, [1983], *On the design of a special purpose scientific programming language*, Softw. Pract. Exp., 13, pp. 893–907.

J. Knight, W. Poole and R. Voigt, [1975], *System balance analysis for vector computers*, Proc. 1975 ACM National Conference, pp. 163–168.

J. Knott, [1983], *A performance analysis of the PASLIB Version 2.1 SEND and RECV routines on the Finite Element Machine*, NASA Contractor Report 172205, NASA Langley Research Center, Hampton, VA.

R. Kober and C. Kuznia, [1978], *SMS—A multiprocessor architecture for high-speed numerical computations*, Proc. 1978 Int. Conf. Par. Proc., pp. 18–23.

P. Kogge, [1973], *Maximal rate pipelines solutions to recurrence problems*, Proc. First Ann. Symp. on Comp. Arch., pp. 71–76.

P. Kogge, [1974], *Parallel solution of recurrence problems*, IBM J. Res. Dev., 18, pp. 138–148.

P. Kogge, [1981], *The Architecture of Pipelined Computers*, McGraw-Hill, New York.

P. Kogge and H. Stone, [1973], *A parallel algorithm for the efficient solution of a general class of recurrence equations*, IEEE Trans. Comput., C-22, pp. 786–793.

V. Konrad and Y. Wallach, [1977], *Iterative solution of linear equations on a parallel processor system*, IEEE Trans. Comput., C-26, pp. 838–847.

H. Kopp, [1977], *Numerical weather forecast with the multi-microprocessor system* SMS201, in Feilmeier [1977], pp. 265–268.

D. Korn and J. Lambiotte, [1979], *Computing the fast Fourier transform on a vector computer*, Math. Comp., 33, pp. 977–992.

J. Kowalik, [1983], *Preliminary experience with multiple-instruction multiple data computation*, in Noor [1983], pp. 49–54.

J. Kowalik, ed., [1984], *Proceedings of the* NATO *Workshop on High Speed Computations*, West Germany, NATO ASI Series, vol. F-7, Springer-Verlag, Berlin.

J. Kowalik and S. Kumar, [1982], *An efficient parallel block conjugate gradient method for linear equations*, Proc. 1982 Int. Conf. Par. Proc., pp. 47–52.

J. Kowalik, S. Kumar, and E. Kamgnia, [1984], *An implementation of the fast-Givens transformations on a* MIMD *computer*, Appl. Math. Polish Academy of Science, to appear.

J. Kowalik, R. Lord and S. Kumar, [1984], *Design and performance of algorithms for* MIMD *parallel computers*, in Kowalik [1984], pp. 257–276.

D. Kuck, [1976], *Parallel processing of ordinary programs*, Advances in Computers 15, Academic Press, New York, pp. 119–179.

D. Kuck, [1977], *A survey of parallel machine organization and programming*, ACM Computing Surveys, 9, pp. 29–59.

D. Kuck, [1978], *The Structure of Computers and Computation*, John Wiley, New York.

D. Kuck, P. Budnick, S. Chen, E. Davis, J. Han, P. Kraska, D. Lawrie, Y. Muraoka, R. Strehendt, and R. Towle, [1973], *Measurement of parallelism in ordinary Fortran programs*, Proc. Sagamore Conf. Parallel Processing, pp. 23–36.

D. Kuck and D. Gajski, [1984], *Parallel processing of sparse structures*, in Kowalik [1984], pp. 229–244.

D. Kuck, D. Lawrie and A. Sameh, eds., [1977], *High Speed Computer and Algorithm Organization*, Academic Press, New York.

D. Kuck, J. McGraw and M. Wolfe, [1984], *A debate: Retire* FORTRAN? Physics Today, 37, pp. 66–75.

D. Kuck, A. Sameh, R. Cytron, A. Veidenbaum, C. Polychronopoulos, G. Lee, T. McDaniel, B. Leasure, C. Beckman, J. Davies, and C. Kruskal, [1984], *The effects of program restructuring, algorithm change and architecture choice on program performance*, Proc. 1984 Int. Conf. Par. Proc., pp. 129–138.

D. Kuck and R. Stokes, [1982], *The Burroughs Scientific Processor* (BSP), IEEE Trans. Comput., C-31, pp. 363–376.

R. Kuhn and D. Padua, [1981], *Parallel Processing*, IEEE Computer Society Press.

A. Kumar, R. Graves, and K. Weilmuenster, [1980], *User's guide for vectorized code* EQUIL *for calculating equilibrium chemistry on the Control Data* STAR-100 *computer*, NASA Tech. Memo. 80192, NASA Langley Research Center, Hampton, VA.

A. Kumar, D. Rudy, J. Drummond, and J. Harris, [1982], *Experiences with explicit finite difference schemes for complex fluid dynamics problems on* STAR-100 *and* CYBER 203 *computers*, in Control Data Corp. [1982].

S. Kumar, [1982], *Parallel algorithms for solving linear equations on* MIMD *computers*, Ph. D. Thesis, Computer Science Dept., Washington State University, Pullman.

S. Kumar and J. Kowalik, [1984], *Parallel factorization of a positive definite matrix on an* MIMD *computer*, Proc. 1984 Int. Conf. Par. Proc., pp. 410–416.

H. Kung, [1976], *Synchronized and asynchronous parallel algorithms for multi-processors*, Algorithms and Complexity, pp. 153–200.

H. Kung, [1979], *Let's design algorithms for* VLSI *systems*, Proc. Conf. Very Large Scale Integration, California Institute of Technology, Pasadena, pp. 65–90.

H. Kung, [1980], *The structure of parallel algorithms*, Advances in Computers 19, M. Yovitts, ed., Academic Press, New York, pp. 65–112.

H. Kung, [1982], *Why systolic architectures?* Computer, 15, 1, pp. 37–46.

H. Kung, [1984], *Systolic algorithms*, in Parter [1984], pp. 127–140.

H. KUNG AND C. LEISERSON, [1979], *Systolic arrays (for VLSI)*, Sparse Matrix Proceedings 1978, I. Duff and G. Stewart, eds., Society for Industrial and Applied Mathematics, Philadelphia, pp. 256–282.

H. KUNG, R. SPROULL AND G. STEELE, eds., [1981], VLSI *Systems and Computations*, Computer Science Press, Rockville, MD.

H. KUNG AND D. STEVENSON, [1977], *A software technique for reducing the routing time on a parallel computer with a fixed interconnection network*, in Kuck, et al. [1977], pp. 423–433.

H. KUNG AND S. YU, [1982], *Integrating high-performance special-purpose devices into a system*, presented at the IBM Symposium on Vector Computers and Scientific Computing, Rome.

S.-Y. KUNG, K. ARUN, D. BHUSKERIO AND Y. HO, [1981], *A matrix data flow language/architecture for parallel matrix operations based on computational wave concept*, in H. Kung, et al [1981].

J. LAMBIOTTE, [1975], *The solution of linear systems of equations on a vector computer*, Ph. D. dissertation, Univ. Virginia, Charlottesville.

J. LAMBIOTTE, [1979], *The development of a STAR-100 code to perform a 2-D FFT.* Proc. Lawrence Livermore Lab. Conf. Sci. Compt., Livermore, CA.

J. LAMBIOTTE, [1983], *Efficient sparse matrix multiplication scheme for the CYBER 203*, in Gary [1984], pp. 243–256.

J. LAMBIOTTE AND L. HOWSER, [1974], *Vectorization on the STAR computer of several numerical methods for a fluid flow problem*, NASA TN D-7545, NASA Langley Research Center, Hampton, VA.

J. LAMBIOTTE AND R. VOIGT, [1975], *The solution of tridiagonal linear systems on the CDC STAR-100 computer*, ACM Trans. Math. Software, 1, pp. 308–329.

J. LARSON, [1984], *Multitasking on the CRAY-X-MP-2 multiprocessor*, Computer, 17, 7, pp. 62–69.

K. LAW, [1982], *Systolic systems for finite element methods*, Dept. of Civil Engineering Report No. R-82-139, Carnegie-Mellon Univ., Pittsburgh.

Lawrence Livermore National Laboratory, [1979], *The S-1 Project*, Lawrence Livermore National Laboratory Report No. UCID-18619, Livermore, CA.

D. LAWRIE, T. LAYMAN, D. BAER, AND J. RANDALL, [1975], *Glypnir—a programming language for Illiac IV*, Comm. ACM, 18, pp. 157–164.

D. LAWRIE AND A. SAMEH, [1983], *Applications of structural mechanics on large-scale multiprocessor computers*, in Noor [1983], pp. 55–64.

D. LAWRIE AND A. SAMEH, [1984], *The computation and communication complexity of a parallel banded system solver*, ACM Trans. Math. Software, 10, pp. 185–195.

P. LECA AND P. ROY, [1983], *Simulation numérique de la turbulence sur un système multi-processor*, First Int. College on Vector and Parallel Methods, Paris.

J. LEE, [1980], *Three-dimensional finite element analysis of layered fiber-reinforced composite materials*, Computers and Structures, 12, p. 319.

R. LEE, [1977], *Performance bounds in parallel processor organizations*, in Kuck, et. al. [1977], pp. 453–455.

M. LEUZE, [1984a], *Parallel triangularization of substructured finite element problems*, ICASE Report No. 84–47, NASA Langley Research Center, Hampton, VA.

M. LEUZE, [1984b], *Parallel triangularization of symmetric sparse matrices by Gaussian elimination*, to appear.

M. LEUZE AND L. SAXTON, [1983], *On minimum parallel computing times for Gaussian elimination*, Congressus Numerantium, 40, pp. 169–179.

J. LEVESQUE AND B. BRODE, [1981], *Efficient Fortran techniques for vector processors*, Pacifica-Sierra Research Corp., Seminar Workbook.

R. LEVINE, [1982], *Supercomputers*, Sci. Amer. 246, January, pp. 118–135.

A. LICHNEWSKY, [1982], *Sur la résolution de systèmes linéares issus de la méthode des élements finis par une multiprocessor*, INRIA Report 119.

A. LICHNEWSKY, [1983], *Some vector and parallel implementations for linear systems arising in PDE problems*, presented at the SIAM Conference on Parallel Processing for Scientific Computing, Norfolk, VA, November.

A. LICHNEWSKY, [1984], *Some vector and parallel implementations for preconditioned conjugate gradient algorithms*, in Kowalik [1984], pp. 343–359.

D. LILES, J. MAHAFFY AND P. GIGUERE, [1984], *An approach to fluid mechanics calculations on serial and parallel computer architectures*, in Parter [1984], pp. 141–160.

N. LINCOLN, [1982], *Technology and design tradeoffs in the creation of a modern supercomputer*, IEEE Trans. Comput., C-31, pp. 349–362.

N. LINCOLN, [1983], *Supercomputers = colossal computations + enormous expectations + renowned risk*, Computer, 16, 5, pp. 38–47.

B. LINDT AND T. AGERWALA, [1981], *Communication issues in the design and analysis of parallel algorithms*, IEEE Trans. Softw. Eng., SE-7, pp. 174–188.

E. LIPITAKIS, [1984], *Solving elliptic boundary value problems on parallel processors by approximate inverse matrix semi-direct methods based on the multiple explicit Jacobi iteration*, Comp. Math., 10, pp. 171–184.

G. LIPOVSKI AND K. DOTY, [1978], *Developments and directions in computer architecture*, Computer, 11, 8, pp. 54–67.

G. LIPOVSKI AND A. TRIPATHI, [1977], *A reconfigurable varistructure array processor*, Proc. 1977 Int. Conf. Par. Proc., pp. 165–174.

J. LIU, [1978], *The solution of mesh equations on a parallel computer*, Dept. Computer Science Report CS-78-19, Waterloo Univ., Waterloo, Ontario.

D. LOGAN, C. MAPLES, D. WEAVER AND W. RATHBUN, [1984], *Adapting scientific programs to the MIDAS multiprocessor system*, Proc. 1984 Int. Conf. Par. Proc., pp. 15–24.

H. LOMAX, [1981], *Some prospects for the future of computational fluid dynamics*, AIAA Comp. Fluid Dyn. Conference, June.

H. LOMAX AND T. PULLIAM, [1982], *A fully implicit factored code for computing three-dimensional flows on the Illiac IV*, in Rodrigue [1982], pp. 217–250.

R. LORD, J. KOWALIK, AND S. KUMAR, [1980], *Solving linear algebraic equations on a MIMD computer*, Proc. 1980 Int. Conf. Par. Proc., pp. 205–210.

R. LORD, J. KOWALIK AND S. KUMAR, [1983], *Solving linear algebraic equations on an MIMD computer*, J. ACM, 30, pp. 103–17.

H. LORIN, [1972], *Parallelism in Hardware and Software*, Prentice-Hall, Englewood Cliffs, NJ.

B. LUBACHEVSKY AND D. MITRA, [1984], *Chaotic parallel computations of fixed points of nonnegative matrices of unit spectral radius*, Proc. 1984 Int. Conf. Par. Proc., pp. 109–16.

F. LUK, [1980], *Computing the singular value decomposition on the Illiac IV*, ACM Trans. Math. Software, 6, pp. 524–539.

S. LUNDSTROM AND G. BARNES, [1980], *A controllable MIMD architecture*, Proc. 1980 Int. Conf. Par. Proc., pp. 19–27.

R. MACCORMACK AND K. STEVENS, [1976], *Fluid dynamics applications of the ILLIAC IV computer*, in Computational Methods and Problems in Aeronautical Fluid Dynamics, Academic Press, New York, pp. 448–465.

N. MADSEN AND G. RODRIGUE, [1975], *Two notes on algorithm design for the CDC STAR-100*, Lawrence Livermore National Laboratory, Tech. Memo. 75-1, Livermore, CA.

N. MADSEN AND G. RODRIGUE, [1976], *A comparison of direct methods for tridiagonal systems on the CDC-STAR-100*, Lawrence Livermore National Laboratory, Preprint UCRL-76993, Rev. 1, Livermore, CA.

N. MADSEN AND G. RODRIGUE, [1977], *Odd-even reduction for pentadiagonal matrices*, in Feilmeister [1977], pp. 103–106.

N. MADSEN, G. RODRIGUE AND J. KARUSH, [1976], *Matrix multiplication by diagonals on a vector parallel processor*, Inf. Proc. Letts., 5, pp. 41–45.

G. MAGÓ, [1979], *A network of multiprocessors to execute reduction languages*, Int. J. Comp. and Info. Sci., 8, pp. 349–385 and 435–471.

G. MAGÓ, [1980], *A cellular computer architecture for functional programming*, Proc. COMPCON Spring, IEEE Comp. Soc. Conf., pp. 179–187.

G. MAGÓ AND R. PARGAS, [1982], *Solving partial differential equations on a cellular tree machine*, Proc. 10th IMACS World Congress on Systems Simulation and Scientific Computation, vol. 1, IMACS, pp. 368–373.

M. MALCOLM AND J. PALMER, [1974], *A fast method for solving a class of tridiagonal linear systems*, Comm. ACM, 17, pp. 14–17

C. MAPLES, D. WEAVER, D. LOGAN AND W. RATHBUN, [1983], *Performance of a modular interactive data analysis system* (MIDAS), Proc. 1983 Int. Conf. Par. Proc., pp. 514–519.

C. MAPLES, D. WEAVER, W. RATHBUN AND D. LOGAN, [1984], *The operation and utilization of the MIDAS multiprocessor architecture*, Proc. 1984 Int. Conf. Par. Proc., pp. 197–206.

H. MARTIN, [1977], *A discourse on a new supercomputer*, PEPE, in Kuck, et al. [1977], pp. 101–112.

C. MCCORMICK, [1982], *Performance of MSC/NASTRAN on the CRAY computer*, in Cray Research, Inc. [1982], pp. 88–98.

L. MCCULLEY AND G. ZAHER, [1974], *Heat shield response to conditions of planetary entry computed on the ILLIAC IV*, unpublished manuscript under NASA Ames Research Center contract 6911.

B. MCDONALD, [1980], *The Chebyshev method for solving nonselfadjoint elliptic equations on a vector computer*, J. Comp. Phys., 35, pp. 147–168.

D. MCGLYNN AND L. SCALES, [1984], *On making the NAG run faster*, in Paddon [1984], pp. 73–89.

J. McGREGOR AND M. SALANA, [1983], *Finite element computation with parallel* VLSI, Proc. 8th ASCE Conf. Elec. Comp., Univ. Houston, Houston, TX, pp. 540–553.

C. MEAD AND L. CONWAY, [1979], *Introduction to* VLSI *Systems*, Addison-Wesley, Reading, MA.

P. MEHROTRA AND T. PRATT, [1982], *Language concepts for distributed processing of large arrays*, Proc. Symp. on Principles of Distributed Computing, Ottawa, Ontario, pp. 19–28.

J. MEIJERINK AND H. VAN DER VORST, [1977], *An iterative solution method for linear systems of which the coefficient matrix is a symmetric M-matrix*, Math. Comp., 31, pp. 148–162.

J. MEIJERINK AND H. VAN DER VORST, [1981], *Guidelines for the usage of incomplete decompositions in solving sets of linear equations as they occur in practical problems*, J. Comp. Phys., 14, pp. 134–155.

R. MELHEM, [1983a], *An abstract systolic model and its application to the design of finite element systems*, Institute for Computational Mathematics and Applications Technical Report No. ICMA-83-66, Univ. Pittsburgh, Pittsburgh.

R. MELHEM, [1983b], *Formal verification of a systolic system for finite element stiffness matrices*, Institute for Computational Mathematics and Applications Technical Report No. ICMA-83-56, Univ. Pittsburgh, Pittsburgh.

R. MELHEM AND W. RHEINBOLDT, [1982], *A mathematical model for the verification of systolic networks*, Institute for Computational Mathematics and Applications Technical Report No. ICMA-82-47, Univ. Pittsburgh, Pittsburgh.

D. MELSON AND J. KELLER, [1983], *Experience in using the* CYBER 203 *and* CYBER 205 *for three-dimensional transonic flow calculations*, AIAA 21st Aerospace Sciences Meeting, January, AIAA Paper 83-0500, also in Control Data Corp. [1982].

R. MENDEZ, [1984], *Benchmark on Japanese-American supercomputers—preliminary results*, IEEE Trans. Comput., C-35, p. 374. An expanded version appeared in the SIAM News 17, 2, March, 1984, p. 3.

M. MERRIAM, [1985], *On the factorization of block-tridiagonals without storage constraints*, SIAM J. Sci. Stat. Comp., 6, pp. 182–192.

G. MEURANT, [1985], *Vector preconditioning for the conjugate gradient method*, to appear.

G. MEYER, [1977], *Effectiveness of multiprocessor networks for solving the nonlinear Poisson equation*, in Kuck, et al. [1977], pp. 323–326.

R. MILLER, [1974], *A comparison of some theoretical models of parallel computation*, IEEE Trans. Comput., C-22, pp. 710–717.

R. MILLSTEIN, [1973], *Control structures in Illiac IV Fortran*, Comm. ACM, 16, pp. 622–627.

M. MINSKY, [1970], *Form and content in computer science*, J. ACM, 17, pp. 197–215.

M. MINSKY AND S. PAPERT, [1971], *On some associative, parallel and analog computations*, Associative Information Techniques, E. Jacks, ed., Elsevier, New York.

W. MIRANKER, [1971], *A survey of parallelism in numerical analysis*, SIAM Rev., 13, pp. 524–547.

W. MIRANKER, [1977], *Parallel methods for solving equations*, in Feilmeister [1977], pp. 9–16.

W. MIRANKER, [1979], *Hierarchical relaxation*, Computing, 23, pp. 267–285.

W. MIRANKER AND W. LINIGER, [1967], *Parallel methods for the numerical integration of ordinary differential equations*, Math. Comp., 21, pp. 303–320.

N. MISSIRLIS AND D. EVANS, [1984], *A second order iterative scheme suitable for parallel implementation*, in Vichnevetsky and Stepleman [1984], pp. 203–206.

K. MIURA, [1971], *The block iterative method for Illiac IV*, Center for Advanced Computation Doc. 41, Univ. Illinois, Urbana.

K. MIURA AND K. UCHIDA, [1984], FACOM *vector processor* VP-100/VP-200 in Kowalik [1984], pp. 127–138.

J. MODI AND M. CLARKE, [1984], *An alternative Givens ordering*, Numer. Math., 43, pp. 83–90.

R. MONTOYE AND D. LAWRIE, [1982], *A practical algorithm for the solution of triangular systems on a parallel processing system*, IEEE Trans. Comput., C-31, pp. 1076–1082.

M. MOORE, R. HIROMOTO AND O. LUBECK, [1985], *Experiences with the Denelcor HEP*, Parallel Computing, 2, to appear.

W. MOORE AND K. STEIGLITZ, [1984], *Efficiency of parallel processing in the solution of Laplace's equation*, in Vichnevetsky and Stepleman [1984], pp. 252–257.

J-M. MORF AND J.-M. DELOSME, [1981], *Matrix decompositions and inversions via elementary signature-orthogonal transformations*, ISSM Int. Symp. Mini & Microcomputers In Control and Measurements, San Francisco.

P. MORICE, [1972], *Calcul parellèle et décomposition dans la résolution d'équations aux derivées partielles de type elliptique*, IRIA, Rocquencourt, France.

M. MORJARIA AND G. MAKINSON, [1984], *Unstructured sparse matrix vector multiplication on the* DAP, in Paddon [1984], pp. 157–166.

T. Moto-oka, ed., [1982], *Fifth Generation Computer Systems*, North-Holland, New York.

T. Moto-oka, [1984], *Japanese project on fifth generation computer systems*, in Kowalik [1984], pp. 99–116.

D. Muller-Wichands and W. Gentzsch, [1982], *Performance comparisons among several parallel and vector computers on a set of fluid flow problems*, DFVLR Report IB 262-82 RO1, Göttingen.

K. Nagel, [1979], *Weather simulation with the multi-microprocessor system* SMS 201, Military Electronics Defense EXPO 78, Proceedings of the Conference, Wisbaden, West Germany, Oct. 3–5, Interario, S. A. Geneva, pp. 60–67.

K. W. Neves, [1984], *Vectorization of scientific software*, in Kowalik [1984], pp. 277–291.

L. Ni and K. Hwang, [1983], *Pipelined evaluation of first-order recurrence systems*, Proc. 1983 Int. Conf. Par. Proc., pp. 537–544.

J. Nievergelt, [1964], *Parallel methods for integrating ordinary differential equations*, Comm. ACM, 7, pp. 731–733.

T. Nodera, [1984], *PCG method for four colored ordered finite difference schemes*, in Vichnevetsky and Stepleman [1984], pp. 222–228.

J. Nolen, D. Kuba and M. Kascic, [1979], *Application of vector processors to the solution of finite difference equations*, Fifth Symposium on Reservoir Simulation, also in SPE J., August 1981.

J. Nolen and P. Stanat, [1981], *Reservoir simulation on vector processing computers*, SPE paper 9649, SPE Middle East Oil Tech. Conf., Manama, Bahrain.

A. Noor, ed., [1983], *Impact of new computing systems on computational mechanics*, American Society of Mech. Engineers.

A. Noor and R. Fulton, [1975], *Impact of the* CDC-STAR-100 *computer on finite-element systems*, J. Structural Div., ASCE 101, no. ST4, pp. 287–296.

A. Noor and S. Hartley, [1978], *Evaluation of element stiffness matrices on* CDC STAR-100 *computer* Computers & Structures, 9, pp. 151–161.

A. Noor, H. Kamel and R. Fulton, [1978], *Substructuring techniques—status and projections*, Computers and Structures, 8, pp. 621–632.

A. Noor and J. Lambiotte, [1978], *Finite element dynamic analysis on the* CDC STAR-100 *computer*, Computers and Structures, 10, pp. 7–19.

A. Noor, O. Storaasli and R. Fulton, [1983], *Impact of new computing systems on finite element computations*, in Noor [1983], pp. 1–32.

A. Noor and S. Voigt, [1975], *Hypermatrix scheme for the* STAR-100 *computer*, Computers and Structures, 5, pp. 287–296.

C. Norrie, [1984], *Supercomputers for superproblems: An architectural overview*, Computer 17, no. 3, pp. 62–74.

D. Norrie, [1984], *The finite element method and large scale computation*, Proc. 4th Int. Symp. on Finite Element Methods in Flow Problems, University of Tokyo Press, Tokyo, North-Holland, Amsterdam, pp. 947–954.

R. Numrich, ed., [1985], *Supercomputer applications symposium*, Proc. Symposium at Purdue Univ., West Lafayette, IN, October 31–November 1, 1984.

W. Oakes and R. Browning, [1979], *Experience running* ADINA *on* CRAY-1, Proc. ADINA Conf., Massachusetts Institute of Technology Report 82448-9, pp. 27–42, Cambridge, MA.

S. O'Donnell, P. Geiger and M. Schultz, [1983], *Solving the Poisson equation on the* FPS-164, Dept. Computer Science Report No. 292, Yale Univ., New Haven, CT.

W. Oed and O. Lange, [1983], *The solution of linear recurrence relations on pipelined processors*, Proc. 1983 Int. Conf. Par. Proc., pp. 545–547.

M. Ogura, M. Sher and J. Ericksen, [1972], *A study of the efficiency of* ILLIAC IV *in hydrodynamic calculations*, Center for Advanced Computation Document No. 59, Univ. Illinois, Urbana-Champaign.

D. O'Leary, [1984], *Ordering schemes for parallel processing of certain mesh problems*, SIAM J. Sci. Stat. Comp., 5, pp. 620–632.

D. O'Leary and G. Stewart, [1984], *Data-flow algorithms for parallel matrix computations*, Computer Science Technical Report No. 1366, Univ. Maryland, College Park.

D. O'Leary and R. White, [1985], *Multi-splittings of matrices and parallel solution of linear systems*, SIAM J. Alg. Disc. Meth., 6, to appear.

D. O'Leary and O. Widlund, [1979], *Capacitance matrix methods for the Helmholtz equation on general three dimensional regions*, Math. Comp. 33, pp. 849–880.

P. Oleinick, [1978], *The implementation of parallel algorithms on an asychronous multiprocessor*, Ph. D. thesis, Dept. Computer Science, Carnegie-Mellon Univ., Pittsburgh.

P. Oleinick, [1982], *Parallel Algorithms on a Multiprocessor*, UMI Research Press.

P. OLEINICK AND S. FULLER, [1978], *The implementation of a parallel algorithm on* C.mmp, Dept. Computer Science Report No. CMU-CS-78-125, Carnegie-Mellon Univ., Pittsburgh.

D. ORBITS, [1978], *A Cray-1 timing simulator*, Systems Engineering Laboratory Report No. 118, Univ. Michigan, Ann. Arbor.

D. ORBITS AND D. CALAHAN, [1976], *Data flow considerations in implementing a full matrix solver with backing store on the* CRAY-1, Systems Engineering Laboratory Report No. 98, Univ. Michigan, Ann. Arbor.

D. ORBITS AND D. CALAHAN, [1978], *A* CRAY-1 *simulator and its application to development of high performance codes*, Proc. LASL Workshop on Vector and Parallel Processors.

S. ORSZAG AND A. PATERA, [1981a], *Subcritical transition to turbulence in planar shear flows*, Transition and Turbulence, R. Meyer, ed., Academic Press, New York, pp. 127–146.

S. ORSZAG AND A. PATERA, [1981b], *Calculation of Von Karman's constant for turbulent channel flow*, Phys. Rev. Lett., 47, pp. 832–835.

S. ORSZAG AND A. PATERA, [1983], *Secondary instability of wall bounded shear flows*, J. Fluid Mech., 128, pp. 347–385.

J. ORTEGA AND R. VOIGT, [1977], *Solution of partial differential equations on vector computers*, Proc. 1977 Army Numerical Analysis and Computers Conference, pp. 475–525.

N. OSTLUND, P. HIBBARD AND R. WHITESIDE, [1982], *A case study in the application of a tightly coupled multi-processor to scientific computations*, in Rodrigue [1982], pp. 375–364.

D. PADDON, [1984], *Supercomputers and Parallel Computation*, Clarendon Press, Oxford.

Y. PAKER, [1977], *Application of microprocessor networks for the solution of equations*, Math. and Comput. Simul., 19, pp. 23–27.

Y. PAKER, [1983], *Multi-Microprocessor Systems*, Academic Press, New York.

J. PALMER, [1974], *Conjugate direction methods and parallel computing*, Ph. D. dissertation, Dept. Computer Science, Stanford Univ., Stanford, CA.

G. PARDEY AND G. THOMAS, [1982], *The implementation of lattice calculations on the* DAP, J. Comp. Phys., 47, pp. 165–178.

R. PARGAS, [1982], *Parallel solution of elliptic partial differential equations on a tree machine*, Ph. D. dissertation, Dept. Computer Science, Univ. North Carolina, Chapel Hill.

D. PARKER, [1980], *Notes on shuffle/exchange type switching networks*, IEEE Trans. Comput., C-29, pp. 213–222.

D. PARKINSON, [1976], *The ICL distributed array processor* DAP, Computational Methods in Classical and Quantum Physics, M. Hooper, ed., Adv. Pub. Ltd.

D. PARKINSON, [1982], *The distributed array processor* (DAP), Comput. Phys. Comm., 28, pp. 325–336.

D. PARKINSON, [1984], *Experience in exploiting large scale parallelism*, in Kowalik [1984], pp. 247–256.

D. PARKINSON AND H. LIDDELL, [1982], *The measurement of performance on a highly parallel system*, IEEE Trans. Comput. C-32, pp. 32–37.

D. PARKINSON AND M. WUNDERLICH, [1984], *A compact algorithm for Gaussian elimination over* $GF(2)$ *implemented on highly parallel computers*, Parallel Computing, 1, pp. 65–73.

S. PARTER, ed., [1984], *Large Scale Scientific Computation*, Academic Press, Orlando, FL.

S. PARTER AND S. STEUERWALT, [1980], *On k-line and* $k \times k$ *block iterative schemes for a problem arising in* 3-D *elliptic difference equations*, SIAM J. Numer. Anal., 17, pp. 823–839.

S. PARTER AND M. STEUERWALT, [1982], *Block iterative methods for elliptic and parabolic difference equations*, SIAM J. Numer. Anal. 19, pp. 1173–1195.

N. PATEL, [1983], *A fully vectorized numerical solution of the incompressible Navier–Stokes equations*, Ph. D. dissertation, Mississippi State Univ., Mississippi State, MS.

N. PATEL AND H. JORDAN, [1985], *A parallelized point rowwise successive over-relaxation method on a multiprocessor*, Parallel Computing, 2, to appear.

G. PAUL, [1985], *Design alternatives of vector processors*, J. Par. and Dist. Comp., 2, to appear.

G. PAUL AND W. WILSON, [1978], *An introduction to* VECTRAN *and its use in scientific applications programming*, Proc. LASL Workshop on Vector and Parallel Processors.

G. PAWLEY AND G. THOMAS, [1982], *The implementation of lattice calculations on the* DAP, J. Comp. Phys., 47, pp. 165–178.

M. PEASE, [1967], *Matrix inversion using parallel processing*, J. ACM, 14, pp. 757–764.

M. PEASE, [1968], *An adaptation of the fast Fourier-transform for parallel processing*, J. ACM, 15, pp. 252–264.

R. PERROTT, [1979], *A standard for supercomputer languages*, in Jesshope and Hockney [1979], pp. 291–308.

F. PETERS, [1984], *Parallel pivoting algorithms for sparse symmetric matrices*, Parallel Computing, 1, pp. 99–110.

V. PETERSON, [1978], *Computational aerodynamics and the NASF*, NASA CR-2032, pp. 5–30.

V. PETERSON, [1984a], *Impact of computers on aerodynamics research and development*, Proc. IEEE, 72, pp. 68–79.

V. PETERSON, [1984b], *Application of supercomputers to computational aerodynamics*, NASA TM-85965, NASA Ames Research Center, Moffett Field, CA.

W. PETERSON, [1983], *Vector Fortran for numerical problems on* CRAY-1, Comm. ACM, 26, pp. 1008–1021.

G. PLATZMAN, [1979], *The* ENIAC *computations of* 1959—*gateway to numerical weather prediction*, Bull. Amer. Meteor. Soc., 60, pp. 302–312.

E. POOLE AND J. ORTEGA, [1985], *Incomplete Choleski conjugate gradient on the* CYBER 203/205, in Numrich [1985].

W. POOLE AND R. VOIGT, [1974], *Numerical algorithms for parallel and vector computers: an annotated bibliography*, Comp. Rev., 15, pp. 379–388.

F. PREPARATA AND D. SARWATE, [1978], *An improved parallel processor bound in fast matrix inversion*, Inf. Proc. Letts., 7, pp. 148–150

F. PREPARATA AND J. VUILLEMIN, [1980], *Optimal integrated circuit implementation of triangular matrix inversion*, Proc. 1980 Int. Conf. Par. Proc., pp. 211–216.

F. PREPARATA AND J. VUILLEMIN, [1981], *The cube-connected cycles: A versatile network for parallel computation*, Comm. ACM, 24, pp. 300–309.

H. PRICE AND K. COATS, [1974], *Direct methods in reservoir simulation*, J. Soc. Pet. Eng., 14, pp. 295–308.

T. PULLIAM AND H. LOMAX, [1979], *Simulation of three-dimensional compressible viscous flow on the Illiac IV computer*, AIAA J., 18, pp. 159–167.

L. PYLE AND S. WHEAT, [1983], *A Kosloff/Basal method* 3D *migration program implemented on the* CYBER 205 *supercomputer*, in Gary [1984], pp. 327–358.

S. RAJAN, [1972], *A parallel algorithm for high-speed subsonic compressible flow over a circular cylinder*, J. Comp. Phys., 12, pp. 534–552.

I. RAJU, [1984], private communication.

I. RAJU AND J. CREWS, [1982], *Three-dimensional analysis of* $[0/90]_s$ *and* $[90/0]_s$ *laminates with a central circular hole*, Composite Tech. Rev., 4, 4, pp. 116–124.

C. RAMAMOORTHY AND H. LI, [1977], *Pipeline architecture*, Comp. Surveys, 9, pp. 61–102.

L. RASKIN, [1978], *Performance evaluation of multiple processor systems*, Dept. Computer Science Report CMU-CS-78-141, Carnegie-Mellon Univ., Pittsburgh.

W. RAY, [1984], *Cyberplus: A multiparallel operating system*, presented at the Los Alamos Workshop on Operating Systems and Environments for Parallel Processing, August 7–9, Los Alamos, NM.

G. REA, [1983], *A software debugging aid for the finite element machine*, Computer Science Dept. Report, Univ. Colorado, Boulder.

S. REDDAWAY, [1979], *The DAP approach*, in Jesshope and Hockney [1979], vol. 2, pp. 309–329.

S. REDDAWAY, [1984], *Distributed Array Processor, architecture and performance*, in Kowalik [1984], pp. 89–98.

D. REDHED, A. CHEN AND S. HOTOVY, [1979], *New approach to* 3D *transonic flow analysis using the* STAR-100 *computer*, AAIA J., 17, pp. 98–99.

D. REED AND M. PATRICK, [1984a], *A model of asynchronous iterative algorithms for solving large sparse linear systems*, Proc. 1984 Int. Conf. Par. Proc., pp. 402–409.

D. REED AND M. PATRICK, [1984b], *Parallel iterative solution of sparse linear systems: models and architectures*, ICASE Report No. 84-35, NASA Langley Research Center, Hampton, VA.

J. REID, [1971], *On the method of conjugate gradients for the solution of large sparse systems of linear equations*, Proc. Conf. Large Sparse Sets of Linear Equations, Academic Press, New York.

J. REID, [1972], *The use of conjugate gradients for systems of linear equations possessing property* A, SIAM J. Numer. Anal., 9, pp. 325–332.

B. REILLY, [1970], *On implementing the Monte Carlo evaluation of the Boltzmann collision integral on* ILLIAC IV, Coordinated Science Laboratory Report No. I-140, Univ. Illinois, Urbana-Champaign.

E. REITER AND G. RODRIGUE, [1984], *An incomplete Choleski factorization by a matrix partition algorithm*, in Birkhoff and Schoenstadt [1984], pp. 161–173.

C. RIEGER, [1981], ZMOB: *hardware from a user's viewpoint*, Proc. IEEE Comput. Soc. Conf. Pattern Recognition and Image Processing, pp. 309–408.

J. RIGANATI AND P. SCHNECK, [1984], *Supercomputing*, Computer, 17, 10, pp. 97–113.

F. ROBERT, [1970], *Méthodes itératives série-parallel*, C. R. Acad. Sci. Paris, 271, pp. 847–850.

F. ROBERT, M. CHARNAY AND F. MUSY, [1975], *Itérations chaotiques série-parallel pour des équations non-linéaires de point fixe*, Appl. Matem., 20, pp. 1–38.

J. ROBINSON, [1979], *Some analysis techniques for asynchronous multiprocessor algorithms*, IEEE Trans. Softw. Eng. SE-5, pp. 24–31.

J. ROBINSON, R. RILEY AND R. HARTKA, [1982], *Evaluation of the SPAR thermal analyzer on the* CYBER-203 *computer*, in Computational Aspects of Heat Transfer and Structures, pp. 405–424.

G. RODRIGUE, ed., [1982], *Parallel Computations*, Academic Press, New York.

G. RODRIGUE, E. GIROUX AND AND M. PRATT, [1980], *Perspectives on large-scale scientific computation*, Computer, 13, 12, pp. 65–80.

G. RODRIGUE, C. HENDRICKSON, AND M. PRATT, [1982], *An implicit numerical solution of the two-dimensional diffusion equation and vectorization experiments*, in Rodrigue [1982], pp. 101–128.

G. RODRIGUE, N. MADSEN AND J. KARUSH, [1976], *Odd-even reduction for banded linear equations*, Lawrence Livermore National Laboratory Report No. UCRL-78652, Livermore, CA.

G. RODRIGUE AND D. WOLITZER, [1984a], *Incomplete block cyclic reduction*, Proc. 10th IMACS World Congress on Systems Simulation and Scientific Computation, vol. 1, IMACS, New Brunswick, NJ, pp. 101–103.

G. RODRIGUE AND D. WOLITZER, [1984b], *Preconditioning by incomplete block cyclic reduction*, Math. Comp., 42, 1984, pp. 549–565.

R. ROGALLO, [1977], *An Illiac program for the numerical simulation of homogeneous incompressible turbulence*, NASA TM-73203, NASA Ames Research Center, Moffett Field, CA.

W. RONSCH, [1984], *Stability aspects in using parallel algorithms*, Parallel Computing, 1, pp. 75–98.

J. ROSENFELD, [1969], *A case study in programming for parallel processors*, Comm. ACM, 12, pp. 645–655.

L. RUDINSKI AND G. PIEPER, [1979], *Evaluating computer program performance on the* CRAY-1, Argonne National Laboratory Report No. 79–9, Argonne, IL.

J. RUDOLPH, [1972], *A production implementation of an associative array processor—-*STARAN, Proc. Fall Joint Comp. Conf., AFIPS Press, Montvale, NJ, pp. 229–241.

T. RUDY, [1980], *Analysis of a 2-D code on the* CRAY-1, Lawrence Livermore National Laboratory Report No. UCID-18549, Livermore, CA.

M. RUSCHITZKU, M. CHONTENSEN, M. AMES, AND B. VICHNEVETSKY, eds., [1984], *Parallel and Large Scale Computers: Performance, Architecture, Applications*, North-Holland, Amsterdam.

R. RUSSELL, [1978], *The* CRAY-1 *computer system*, Comm. ACM, 21, pp. 63–72.

Y. SAAD, [1983a], *Least squares polynomials in the complex plane with applications to solving sparse non-symmetric matrix problems*, Dept. Computer Science Report No. RR-276, Yale Univ., New Haven, CT.

Y. SAAD, [1983b], *Practical use of polynomial preconditionings for the conjugate gradient method*, Computer Science Report No. RR-282, Yale Univ., New Haven, CT.

Y. SAAD AND A. SAMEH, [1981a], *A parallel block Stiefel method for solving positive definite systems*, in Schultz [1981], pp. 405–411.

Y. SAAD AND A. SAMEH, [1981b], *Iterative methods for the solution of elliptic difference equations on multiprocessors*, CONPAR 81, pp. 395–411.

Y. SAAD, A. SAMEH AND P. SAYLOR, [1985], *Parallel iterative methods for elliptic difference equations*, SIAM J. Sci. Stat. Comp., 6, to appear.

M. SALAMA, S. UTKU, AND R. MELOSH, [1983], *Parallel solution of finite element equations*, Proc. 8th ASCE Conf. Elec. Comp., University of Houston, pp. 526–539.

A. SAMEH, [1971a], *Illiac IV applications*, Proc. 9th Annual Allerton Conf. Circuit System Theory, pp. 1030–1038.

A. SAMEH, [1971b], *On Jacobi and Jacobi-like algorithms for a parallel computer*, Math. Comp., 25, pp. 579–590.

A. SAMEH, [1977], *Numerical parallel algorithms—a survey*, in Kuck, et al. [1977], pp. 207–228.

A. SAMEH, [1981], *Parallel algorithms in numerical linear algebra*, presented at the CREST Conference.

A. SAMEH, [1982], *Solving the linear least squares problem on a linear array of processors*, Proc. Purdue Workshop on Algorithmically-Specialized Computer Organizations, Academic Press, New York.

A. SAMEH, [1983], *An overview of parallel algorithms in numerical linear algebra*, EDF-Bull. des Etudes et des Rech. Ser C, 1, pp. 129–134.

A. SAMEH, [1984a], *A fast Poisson solver for multiprocessors*, in Birkhoff and Schoenstadt [1984], pp. 175–186.

A. SAMEH, [1984b], *On two numerical algorithms for multiprocessors*, in Kowalik [1984], pp. 311–328.

A. SAMEH AND R. BRENT, [1977], *Solving triangular systems on a parallel computer*, SIAM J. Numer. Anal., 14, pp. 1101–1113.

A. SAMEH, S. CHEN, AND D. KUCK, [1976], *Parallel Poisson and biharmonic solvers*, Computing, 17, pp. 219–230.

A. SAMEH AND D. KUCK, [1977a], *A parallel QR algorithm for symmetric tridiagonal matrices*, IEEE Trans. Comput., C-26, pp. 147–153.

A. SAMEH AND D. KUCK, [1977b], *Parallel direct linear system solvers-a survey*, in Feilmeister [1977], pp. 25–30.

A. SAMEH AND D. KUCK, [1978], *On stable parallel linear system solvers*, J. ACM, 25, pp. 81–91.

A. SAMEH AND C. TAFT, [1982], *Preconditioning strategies for the conjugate gradient algorithm on multi-processors*, Presented at the 1982 Sparse Matrix Symposium.

V. SAUNDERS AND M. GUEST, [1982], *Applications of the Cray-1 for quantum chemistry calculations*, Comp. Phys. Comm., 26, pp. 389–395.

A. SAWCHUK AND T. STRAND, [1984], *Digital optical computing*, Proc. IEEE, 72, pp. 758–779.

U. SCHNENDEL, [1984], *Introduction to Numerical Methods for Parallel Computers*, B. W. Conolly, Trans., Halsted Press, New York.

E. SCHNEPF AND W. SCHONAUER, [1983], *Parallelization of PDE software for vector computers*, Proc. Parallel Computing 83, Berlin.

W. SCHONAUER, [1983a], *The efficient solution of large linear systems resulting from the FDM for 3-D PDE's on vector computers*, Proc. First. Intern. Coll. on Vector and Parallel Computing in Scientific Applications, A. Bassanut, ed., Bull. de la Direction des Etudes et Recherches, Ser. C., 1, pp. 135–142.

W. SCHONAUER, [1983b], *Numerical experiments with instationary Jacobi-QR methods for the iterative solution of linear equations*, ZAMM, 63, pp. T380–T382.

W. SCHONAUER AND K. RAITH, [1982], *A polyalgorithm with diagonal storing for the solution of very large indefinite linear banded systems on a vector computer*, Proc. 10th IMACS World Congress on Systems Simulation and Scientific Computation, vol. 1, IMACS, New Brunswick, NJ, pp. 326–328.

W. SCHONAUER, E. SCHNEPF AND H. MULLER, [1984], *PDE software for vector computers*, in Vichnevetsky and Stepleman [1984], pp. 258–267.

W. SCHONAUER, E. SCHNEPF, AND K. RAITH, [1983], *The redesign and vectorization of the SLDGL-program package for the self-adaptive solution of nonlinear systems of elliptic and parabolic PDE's.*, Conference of the IFFP Working Group 2.5 on Numerical Software, Sweden.

W. SCHONAUER, E. SCHNEPF, AND K. RAITH, [1984], *Modularization of PDE software for vector computers*, ZAMM, 64, pp. T309–T312.

R. SCHREIBER, [1984], *Systolic arrays: High performance parallel machines for matrix computation*, in Birkhoff and Schoenstadt [1984], pp. 187–194.

R. SCHREIBER AND P. KUEKES, [1982], *Systolic linear algebra machines in digital signal processing*, Proc. USC Workshop on VLSI and Modern Signal Processing, Los Angeles, Prentice-Hall, Englewood Cliffs, NJ.

R. SCHREIBER AND W. TANG, [1982], *Vectorizing the conjugate gradient method*, in Control Data Corp. [1982].

M. SCHULTZ, ed., [1981], *Elliptic Problem Solvers*, Academic Press, New York.

M. SCHULTZ, [1984], *Solving elliptic problems on an array processor system*, in Birkhoff and Schoenstadt [1984], pp. 77–92.

J. SCHWARTZ, [1980], *Ultracomputers*, ACM Trans. Program. Lang. Syst., 2, pp. 484–521.

J. SCHWARTZ, [1983], *A taxonomic table of parallel computers based on 55 designs*, Ultracomputer Note No. 69, Courant Institute, New York Univ., New York.

R. SCOTT, [1981], *On the choice of discretization for solving PDE's on a multi-processor*, in Schultz [1981], pp. 419–422.

C. SEITZ, [1982], *Ensemble architectures for VLSI—a survey and taxonomy*, Proc. MIT Conf. on Advanced Res. in VLSI, Artech Books, pp. 130–135.

C. SEITZ, [1984], *Experiments with VLSI ensemble machines*, J. VLSI and Comp. Sys., to appear.

C. SEITZ AND J. MATISOO, [1984], *Engineering limits on computer performance*, Physics Today, 37, 5, pp. 38–45.

M. SEJNOWSKI, E. UPCHURCH, R. KAPUR, D. CHARLU AND G. LIPOVSKI, [1980], *An overview of the Texas reconfigurable array computer*, AFIPS Conf. Proc., 1980, NCC, pp. 631–641.

A. SHAH, [1980], *Group broadcast mode of interprocessor communications for the finite element machine*, Dept. of Computer Science Report CSDG-80-1, Univ. Colorado, Boulder.

J. SHANEHCHI AND D. EVANS, [1981], *New variants of the quadrant interlocking factorization (QIF) method*, CONPAR 81 Conf. Proc. Lecture Notes in Computer Science III, W. Handler, ed., Springer-Verlag, Berlin, pp. 493–507.

J. SHANEHCHI AND D. EVANS, [1982], *Further analysis of the QIF method*, Int. J. Comput. Math., 11, pp. 143–154.

J. SHANG, P. BUNING, W. HANKEY AND M. WIRTH, [1980], *Performance of a vectorized three-dimensional Navier–Stokes code on the CRAY-1 computer*, AIAA J., 18, pp. 1073–1079.

D. SHAW, [1984], *SIMD and MSIMD variants of the NON-VON supercomputer*, Proc. COMPCON 84, IEEE Comp. Soc. Conf., pp. 360–363.

G. SHEDLER, [1967], *Parallel numerical methods for the solution of equations*, Comm. ACM, 10, pp. 286–291.

T. SHIMADA, K. HIRAKI AND K. NISHIDA, [1984], *An architecture of a data flow computer and its evaluation*, Proc. COMPCON 84, IEEE Comp. Soc. Conf., pp. 486–490.

H. SIEGEL, [1979], *Intercommunication networks for SIMD machines*, Computer, 12, 6, pp. 57–65.

L. SIEGEL, H. SIEGEL, AND P. SWAIN, [1982], *Performance measurements for evaluating algorithms for SIMD machines*, IEEE Trans. Soft. Eng., SE-8, pp. 319–331.

D. SIEWIOREK, [1983], *State of the art in parallel computing*, in Noor [1983], pp. 33–48.

D. SLOTNICK, W. BORCK AND R. MCREYNOLDS, [1962], *The SOLOMON computer*, Proc. AFIPS, FJCC, 22, pp. 97–107.

B. SMITH, [1978], *A pipelined, shared resource MIMD computer*, Proc. 1978 Int. Conf. Par. Proc., pp. 6–8.

R. SMITH AND J. PITTS, [1979], *The solution of the three-dimensional viscous compressible Navier–Stokes equations on a vector computer*, Advances in Computer Methods for Partial Differential Equations-III, IMACS, New Brunswick, NJ, pp. 245–252.

R. SMITH, J. PITTS AND J. LAMBIOTTE, [1978], *A vectorization of the Jameson–Caughey NYU transonic swept-wing computer program FLO-22-VI for the STAR-100 computer*, NASA TM-78665, NASA Langley Research Center, Hampton, VA.

L. SNYDER, [1982], *Introduction to the configurable highly parallel computer*, Computer, 15, 1, pp. 47–56.

P. SOLL, N. HABRA AND G. RUSSEL, [1977], *Experience with a vectorized general circulation climate model on STAR-100*, in Kuck, et al. [1977], pp. 311–312.

M. SOLOMON AND R. FINKEL, [1979], *The Roscoe operating system*, Proc. 7th Symp. Op. Sys. Princ., pp. 108–114.

D. SORENSEN, [1984], *Buffering for vector performance on a pipelined MIMD machine*, Argonne National Laboratory Report No. ANL/MCS-TM–29, Argonne, IL.

D. SORENSEN, [1985], *Analysis of pairwise pivoting in Gaussian elimination*, IEEE Trans. Comput., TC-34, pp. 274–279.

J. SOUTH, J. KELLER AND M. HAFEZ, [1980a], *Computational transonics on a vector computer*, U. S. Army Numerical Analysis and Computers Conference, ARO Rep. No. 80-3, August, pp. 357–368.

J. SOUTH, J. KELLER AND M. HAFEZ, [1980b], *Vector processor algorithms for transonic flow calculations*, AIAA J., 18, pp. 786–792.

M. SRINIVAS, [1983], *Optimal parallel scheduling of Gaussian elimination DAG's*, IEEE Trans. Comput., C-32, pp. 1109–1117.

P. STANAT AND J. NOLEN, [1982], *Performance comparisons for reservoir simulation problems on three supercomputers*, 6th SPE Symposium Reservoir Simulation, also in Control Data Corp. [1982].

K. STEVENS, [1975], *CFD-A Fortran-like language for the Illiac IV*, Sigplan Notices, pp. 72–80.

K. STEVENS, [1979], *Numerical aerodynamics simulation facility project*, in Jesshope and Hockney [1979], vol. 2, pp. 331–342.

H. STONE, [1971], *Parallel processing with the perfect shuffle*, IEEE Trans. Comput., C-20, pp. 153–161.

H. STONE, [1973], *An efficient parallel algorithm for the solution of a tridiagonal linear system of equations*, J. ACM, 20, pp. 27–38.

H. STONE, [1975], *Parallel tridiagonal equation solvers*, ACM Trans. Math Software, 1, pp. 289–307.

H. STONE, [1980], *Parallel computation*, in Introduction to Computer Architecture, Second Edition, H. Stone, ed., Science Research Associates, Inc., pp. 363–425.

O. STORAASLI, S. PEEBLES, T. CROCKETT, J. KNOTT AND L. ADAMS, [1982], *The finite element machine: An experiment in parallel processing*, Proc. Conf. on Res. in Structures and Solid Mech., NASA Conf. Pub. 2245, NASA Langley Research Center, Hampton, VA, pp. 201–217.

J. STRIKWERDA, [1982], *A time split difference scheme for the compressible Navier–Stokes equations with applications to flows in slotted nozzles*, in Rodrigue [1982], pp. 251–267.

J. STRINGER, [1982], *Efficiency of D4 Gaussian elimination on a vector computer*, in Cray Research, Inc. [1982], pp. 115–121.

H. SULLIVAN AND T. BASHKOW, [1977], *A large scale homogeneous fully distributed parallel machine*, Proc. 4th Annual Symp. Comp. Arch., pp. 105–117.

R. SWAN, S. FULLER AND D. SIEWIOREK, [1977], *Cm* - a modular multimicroprocessor*, Proc. AFIPS Nat. Computer Conf., AFIPS Press, Montvale, NJ, pp. 637–644.

P. SWARZTRAUBER, [1977], *The methods of cyclic reduction, Fourier analysis and the FACR algorithm for the discrete solution of Poisson's equation on a rectangle*, SIAM Rev., 19, pp. 490–501.

P. SWARZTRAUBER, [1979a], *A parallel algorithm for solving general tridiagonal equations*, Math. Comp. 33, pp. 185–199.

P. SWARZTRAUBER, [1979b], *The solution of tridiagonal systems on the CRAY-1*, in Jesshope and Hockney [1979], vol. 2, pp. 343–358.

P. SWARZTRAUBER, [1982], *Vectorizing the FFTs*, in Rodrigue [1982], pp. 51–83.

P. SWARZTRAUBER, [1983], *Efficient algorithms for pipeline and parallel computers*, in Noor [1983], pp. 89–104.

P. SWARZTRAUBER, [1984], *FFT algorithms for vector computers*, Parallel Computing, 1, pp. 45–63.

C. TAFT, [1982], *Preconditioning strategies for solving elliptic equations on a multiprocessor*, Computer Science Dept. Report, Univ. Illinois, Urbana-Champaign.

Y. TAKAHASHI, [1982], *Partitioning and allocation in parallel computation of partial differential equations*, Proc. 10th IMACS World Congress on Systems Simulation and Scientific Computation, vol. 1, IMACS, New Brunswick, NJ, pp. 311–313.

C. TEMPERTON, [1979a], *Direct methods for the solution of the discrete Poisson equation: some comparisons*, J. Comp. Phys., 31, pp. 1–20.

C. TEMPERTON, [1979b], *Fast Fourier transforms and Poisson solvers on* CRAY-1, in Jesshope and Hockney, [1979], vol. 2, pp. 359–379.

C. TEMPERTON, [1979c], *Fast Fourier transforms on* CRAY-1, European Center for Medium Range Weather Forecasts Report No. 21.

C. TEMPERTON, [1980], *On the* FACR(l) *algorithm for the discrete Poisson equation*, J. Comp. Phys., 34, pp. 314–329.

C. TEMPERTON, [1984], *Fast Fourier transforms on the* CYBER 205, in Kowalik [1984], pp. 403–416.

G. TENNILLE, [1982], *Development of a one-dimensional stratospheric analysis program for the* CYBER 203, in Control Data Corp. [1982].

W. THOMPKINS AND R. HAIMES, [1983], *A minicomputer/array processor/memory system for large scale fluid dynamic calculations*, in Noor [1983], pp. 117–126.

K. THURBER, [1976], *Large Scale Computer Architectures: Parallel and Associative Processors*, Hayden Book Co.,

K. THURBER AND L. WALD, [1975], *Associative and parallel processors*, Comput. Surveys, 7, pp. 215–245.

J. TIBERGHIEN, ed., [1984], *New Computer Architectures*, Academic Press, Orlando, FL.

D. TOLLE AND W. SIDDALL, [1981], *On the complexity of vector computations in binary tree machines*, Inf. Proc. Lett., 13, pp. 120–124.

J. TRAUB, ed., [1974a], *Complexity of Sequential and Parallel Numerical Algorithms*, Academic Press, New York.

J. TRAUB, [1974b], *Iterative solution of tridiagonal systems on parallel or vector computers*, in Traub, [1974a], pp. 49–82.

P. TRELEAVEN, [1979], *Exploiting program concurrency in computing systems*, Computer, 12, 1, pp. 42–50.

P. TRELEAVEN, [1984], *Decentralised computer architecture*, in Tiberghien, [1984], pp. 1–58.

L. UHR, [1984], *Algorithm Structured Computer Arrays and Networks*, Academic Press, New York, 1984.

S. UNGER, [1958], *A computer oriented towards spatial problems*, Proc. IRE, 46, pp. 1744–1750.

M. VAJTERSIC, [1979], *A fast parallel method for solving the biharmonic boundary value problem on a rectangle*, Proc. First European Conference on Parallel Distributed Processing, Toulouse, pp. 136–141.

M. VAJTERSIC, [1981], *Solving two modified discrete Poisson equations in* $7 \log N$ *steps on* N^2 *processors*, CONPAR81, pp. 423–432.

M. VAJTERSIC, [1982], *Parallel Poisson and biharmonic solvers implemented on the EGPA multiprocessor*, Proc. 1982 Int. Conf. Par. Proc. pp. 72–81.

H. VAN DER VORST, [1981], *A vectorizable variant of some ICCG methods*, SIAM J. Sci. Stat. Comput., 3, pp. 350–356.

H. VAN DER VORST, [1983], *On the vectorization of some simple ICCG methods*, First Int. Conf. Vector and Parallel Computation in Scientific Applications, Paris, 1983.

J. VAN ROSENDALE, [1983a], *Algorithms and data structures for adaptive multigrid elliptic solvers*, Appl. Math. Comp. 13, pp. 453–470.

J. VAN ROSENDALE, [1983b], *Minimizing inner product data dependencies in conjugate gradient iteration*, Proc. 1983 Int. Conf. Par. Proc., pp. 44–46.

F. VAN SCOY, [1977], *Some parallel cellular matrix algorithms*, Proc. ACM Comp. Sci. Conf.

R. VARGA, [1962], *Matrix Iterative Analysis*, Prentice Hall, Englewood Cliffs, NJ.

V. VENKAYYA, D. CALAHAN, P. SUMMERS AND V. TISCHLER, [1983], *Structural optimization on vector processors*, in Noor [1983], pp. 155–190.

R. VICHNEVETSKY AND R. STEPLEMAN, eds., [1984], *Advances in computer methods for partial differential equations,-V*, Proc. Fifth IMACS International Symposium, Lehigh Univ., Bethlehem, PA, June, 1984.

R. VOIGT, [1977], *The influence of vector computer architecture on numerical algorithms*, in Kuck, et al. [1977], pp. 229–244.

R. VOIGT, D. GOTTLIEB AND M. HUSSAINI, eds., [1984], *Spectral Methods for Partial Differential Equations*, Society for Industrial and Applied Mathematics, Philadelphia.

R. VOITUS, [1981], *A multiple process software package for the finite element machine*, Computer Science Dept. Report, Univ. Colorado, Boulder,

J. VON NEUMANN, [1966], *A system of 29 states with a general transition rule*, Theory of Self-Reproducing Automata, A. Burks, ed., Univ. Illinois Press, Urbana, Champaign, pp. 305–317.

D. VRSALOVIC, D. SIEWIOREK, A. SEGALL AND E. GEHRINGER, [1984], *Performance prediction for multi-processor systems*, Proc. 1984 Int. Conf. Par. Proc., pp. 139–146.

R. WAGNER, [1983], *The Boolean vector machine*, 1983 IEEE Conference Proc. 10th Annual Int. Symp. Comp. Arch., pp. 59–66.

R. WAGNER, [1984], *Parallel solution of arbitrarily sparse linear systems*, Dept. Computer Science Report No. CS-1984-13, Duke Univ., Durham, NC.

Y. WALLACH AND V. KONRAD, [1976], *Parallel solutions of load flow problems*, Arch. Elektrotechnik, 57, pp. 345–354.

Y. WALLACH AND V. KONRAD, [1980], *On block parallel methods for solving linear equations*, IEEE Trans. Comput. C-29, pp. 354–359.

J. WALLIS AND J. GRISHAM, [1982], *Reservoir simulation on the* CRAY-1, in Cray Research, Inc. [1982], pp. 122–139.

J. WALLIS AND J. GRISHAM, [1982], *Petroleum reservoir simulation on* CRAY-1 *and on the* FPS-164, in Proc. 10th IMACS World Congress on Systems Simulation and Scientific Computation, vol. 1, IMACS, New Brunswick, NJ, pp. 308–310.

H. WANG, [1981], *A parallel method for tridiagonal equations*, ACM Trans. Math. Software, 7, pp. 170–183.

H. WANG, [1982a], *On vectorizing the fast Fourier transform*, BIT, 20, pp. 233–243.

H. WANG, [1982b], *Vectorization of a class of preconditioned conjugate gradient methods for elliptic difference equations*, IBM Scientific Center, Palo Alto, CA.

W. WARE, [1973], *The ultimate computer*, IEEE Spect., 10, 3, pp. 89–91.

P. WATANABE, J. FLOOD AND S. YEN, [1974], *Implementation of finite difference schemes of solving fluid dynamic problems on Illiac IV*, Coordinated Science Laboratory Report No. T-11, Univ. Illinois, Urbana, Champaign.

I. WATSON AND J. GURD, [1982], *A practical data flow computer*, Computer, 15, 2, pp. 51–57.

W. WATSON, [1972], *The TI-ASC, A highly modular and flexible super computer architecture*, Proc. AFIPS, 41, pt. 1, pp. 221–228.

J. WATTS, [1979], *A conjugate gradient truncated direct method for the iterative solution of the reservoir simulation pressure equation*, Proc. SPE 54th Annual Fall Technical Conference and Exhibition, Las Vegas, NV.

S. WEBB, [1980], *Solution of partial differential equations on the* ICL *distributed array processor*, ICL Technical J., pp. 175–190.

S. WEBB, J. MCKEONN AND D. HUNT, [1982], *The solution of linear equations on a* SIMD *computer using a parallel iterative algorithm*, Comp. Phys. Comm. 26, pp. 325–329.

E. WEIDNER AND J. DRUMMOND, [1982], *Numerical study of staged fuel injection for supersonic combustion*, AIAA J., 20, pp. 1426–1431.

J. WEILMUNSTER AND L. HOWSER, [1976], *Solution of a large hydrodynamic problem using the* STAR-100 *computer*, NASA TM X-73904, NASA Langley Research Center, Hampton, VA.

J. WELSH, [1982], *Geophysical fluid simulation on a parallel computer*, in Rodrigue [1982], pp. 269–277.

O. WIDLUND, [1984], *Iterative methods for elliptic problems on regions partitioned into substructures and the biharmonic Dirichlet problem*, Dept. Computer Science Report 101, Courant Institute, New York Univ., NY.

R. WILHELMSON, [1974], *Solving partial differential equations using* ILLIAC IV, in Constructive and Computational Methods for Differential and Integral Equations, A. Dold and B. Eckmann, eds., Springer-Verlag, New York, pp. 453–476.

J. WILKINSON, [1954], *The calculation of the latent roots and vectors of matrices on the pilot model of the* ACE, Proc. Camb. Phil. Soc. 50, Pt. 4, pp. 536–566.

S. WILLIAMS, [1979], *The portability of programs and languages for vector and array processors*, in Jesshope and Hockney [1979], vol. 2, pp. 381–94.

D. WILLIAMSON, [1983], *Computational aspects of numerical weather prediction on the* CRAY *computer*, in Noor [1983], pp. 127–140.

D. WILLIAMSON AND P. SWARZTRAUBER, [1984], *A numerical weather prediction model - computational aspects*, Proc. IEEE, 72, pp. 56–67.

E. WILSON, [1983], *Finite element analysis on microcomputers*, in Noor [1983], pp. 105–116.

K. WILSON, [1982], *Experience with an FPS array processor*, in Rodrigue [1982], pp. 279–314.

O. WING AND J. HUANG, [1977], *A parallel triangulation process of sparse matrices*, Proc. 1977 Int. Conf. Par. Proc., pp. 207–214.

O. WING AND J. HUANG, [1980], *A computational model of parallel solutions of linear equations*, IEEE Trans. Comput., TC-29, pp. 632–638.

N. WINSOR, [1981], *Vectorization of fluid codes*, in Finite Difference Techniques for Vectorized Fluid Dynamics Calculations, D. Book, ed., Springer-Verlag, New York, NY, pp. 152–163.

L. WITTIE, [1980], *Architectures for large networks of microcomputers*, Workshop in Interconnection Networks for Parallel and Distributed Processing, April, pp. 31–40.

L. WITTIE AND A. VAN TILBOUG, [1980], *Micros, a distributed operating system for micronet, a reconfigurable network computer*, IEEE Trans. Comp., C-29, pp. 1133–1144.

P. WOODWARD, [1982], *Trade-offs in designing explicit hydrodynamic schemes for vector computers*, in Rodrigue [1982], pp. 153–171.

J. WORLTON, [1981], *A philosophy of supercomputing*, Los Alamos National Laboratory Report No. LA-8849-MS, Los Alamos, NM.

J. WORLTON, [1984], *Understanding supercomputer benchmarks*, Datamation, 30, 14, pp. 121–130.

C. WU, J. FERZIGER, D. CHAPMAN AND R. ROGALLO, [1983], *Navier-Stokes simulation of homogeneous turbulence on the CYBER 205*, in Gary [1984], pp. 227–239.

W. WULF AND C. BELL, [1972], C.mmp—*a multiminiprocessor*, Proc. AFIPS Fall Joint Comp. Conf., AFIPS Press, Reston, VA, pp. 765–777.

W. WULF AND S. HARBISON, [1978], *Reflections in a pool of processors*, Dept. Computer Science Technical Report, Carnegie-Mellon Univ., Pittsburgh.

M. YASUMURA, Y. TANAKA AND Y. KANADA, [1984], *Compiling algorithms and techniques for the S-810 vector processor*, Proc. 1984 Int. Conf. Par. Proc., pp. 285–290.

D. YOUNG, [1971], *Iterative Solution of Large Linear Systems*, Academic Press, New York.

N. Y. YOUSIF, [1983], *Parallel algorithms for asynchronous multiprocessors*, Ph.D. Thesis, Loughborough Univ., England.

N. YU AND P. RUBBERT, [1982], *Transonic flow simulations for 3D complex configurations*, in Cray Research, Inc. [1982], pp. 41–47.

V. ZAKHAROV, [1984], *Parallelism and array processing*, IEEE Trans. Comput., C-33, pp. 45–78.

P. ZAVE AND G. COLE, [1983], *A quantitative evaluation of the feasibility of and suitable hardware structures for an adaptive parallel finite element system*, ACM Trans. Math. Software 9, pp. 271–292.

P. ZAVE AND W. RHEINBOLDT, [1979], *Design of an adaptive parallel finite element system*, ACM Trans. Math. Software, 5, pp. 1–17.

INDEX

93